MEMOIRS
of the
American Mathematical Society

ISSN 0065-9266 (print)
ISSN 1947-6221 (online)

Number 1500

Exponential Decay Estimates and Smoothness of the Moduli Space of Pseudoholomorphic Curves

Kenji Fukaya
Yong-Geun Oh
Hiroshi Ohta
Kaoru Ono

July 2024 • Volume 299 • Number 1500 (seventh of 7 numbers)

AMS AMERICAN MATHEMATICAL SOCIETY

Library of Congress Cataloging-in-Publication Data

Cataloging-in-Publication Data has been applied for by the AMS.
See http://www.loc.gov/publish/cip/.
DOI: https://doi.org/10.1090/memo/1500

Memoirs of the American Mathematical Society

This journal is devoted entirely to research in pure and applied mathematics.

Subscription information. Beginning in 2024, Memoirs will be published monthly through 2026. Memoirs is also accessible from www.ams.org/journals. The 2024 subscription begins with volume 293 and consists of twelve mailings, each containing one or more numbers. Individual subscription prices for 2024 are as follows. For electronic only: US$1013. For paper delivery: US$1,163. Add US$22 for delivery within the United States; US$150 for surface delivery outside the United States. Upon request, subscribers to paper delivery of this journal are also entitled to receive electronic delivery. For information on institutional pricing, please visit `https://www.ams.org/publications/journals/subscriberinfo`. Subscription renewals are subject to late fees. See `www.ams.org/journal-faq` for more journal subscription information. Each number may be ordered separately; please specify number when ordering an individual number.

Back number information. For back issues see `www.ams.org/backvols`.

Subscriptions and orders should be addressed to the American Mathematical Society, P. O. Box 845904, Boston, MA 02284-5904 USA. *All orders must be accompanied by payment.* Other correspondence should be addressed to 201 Charles Street, Providence, RI 02904-2213 USA.

Memoirs of the American Mathematical Society (ISSN 0065-9266 (print); 1947-6221 (online)) is published bimonthly (each volume consisting usually of more than one number) by the American Mathematical Society at 201 Charles Street, Providence, RI 02904-2213 USA. Periodicals postage paid at Providence, RI. Postmaster: Send address changes to Memoirs, American Mathematical Society, 201 Charles Street, Providence, RI 02904-2213 USA.

This publication is indexed in *Mathematical Reviews®*, *Zentralblatt MATH*, *Science Citation Index®*, *Science Citation Index™-Expanded*, *ISI Alerting Services℠*, *SciSearch®*, *Research Alert®*, *CompuMath Citation Index®*, *Current Contents®/Physical, Chemical & Earth Sciences*.
This publication is archived in *Portico* and *CLOCKSS*.
Printed in the United States of America.

∞ The paper used in this book is acid-free and falls within the guidelines
established to ensure permanence and durability.
Visit the AMS home page at `https://www.ams.org/`

10 9 8 7 6 5 4 3 2 1 29 28 27 26 25 24

Contents

Abstract

The compacified moduli space of bordered stable maps carries a Kuranishi structure with boundary. Smoothness of Kuranishi structure along the boundary requires smoothness of coordinate changes along the boundary. The proof of smoothness is written by the authors in [AMS/IP Stud. Adv. Math., 46.1, 2009, xii+396 pp.], [AMS/IP Stud. Adv. math., 46.2, 2009, pp. i–xii and 397–805], and [Surv. Differ. Geom., vol. 22, pp. 133–190, 2018] based on some uniform exponential decay estimates of the stable maps with respect to a parameter T, the length of the gluing cylindrical neck, near the boundary of the moduli space. In this paper we establish this exponential decay estimates in a precise manner by carefully examining the dependence on the parameter T of the gluing construction in the setting of bordered Riemann surfaces with boundary punctures. We also show that the smoothness of the collar follows from the aforementioned exponential estimates by taking $s = 1/T$ as the radial coordinate of the collar.

Received by the editor March 22, 2016, and, in revised form, May 7, 2018, and March 9, 2021.

Article electronically published on July 11, 2024.

DOI: https://doi.org/10.1090/memo/1500

2020 *Mathematics Subject Classification.* Primary 53D35, 58D27, 35B40; Secondary 53D40, 53D45.

Key words and phrases. pseudo-holomorphic curve, Kuranishi structure, Gluing, Exponential decay, Weighted Sobolev space.

The first author was partially supported by JSPS Grant-in-Aid for Scientific Research No. 23224002, NSF grant # 1406423 and Simons Collaboration on homological Mirror symmetry.

The second author was partially supported by IBS project # IBS-R003-D1 in Korea.

The third author was partially supported by JSPS Grant-in-Aid for Scientific Research Nos.19340017, 23340015, 15H02054, 21H00983, 21K18576, 21H00985, 23K20796.

The fourth author was partially supported by JSPS Grant-in-Aid for Scientific Research, Nos. 18340014, 21244002, 26247006, 19H00636.

CHAPTER 1

Introduction

The theory of Kuranishi structures is introduced by the first and the fourth named authors in [**FOn**], and further amplified by the authors in [**FOOO1**], [**FOOO2**], [**FOOO4**]. To implement the abstract theory of Kuranishi structure in the study of concrete moduli problem, especially that of the moduli space of pseudoholomorphic curves, one issue is to establish smoothness of the Kuranishi map $\mathfrak{s}$ and of the coordinate change of the Kuranishi neighborhoods.

Let us first describe the problem in a bit more detail in the context of bordered stable pseudoholomorphic curves attached to Lagrangian submanifolds.[1] Note that an object representing a point near the boundary of our moduli space contains a neck region, which is a long strip $[-T,T]\times[0,1]$, and the boundary point corresponds to the case where the source curve is singular, that is, the case when $T=\infty$.

Note ∞ is included in $(T_0,\infty]$. As a topological space $(T_0,\infty]$ has unambiguous meaning. On the other hand there is no obvious choice of its smooth structure as a manifold with boundary. Moreover for several relevant maps such as the Kuranishi map in the definition of Kuranishi structure, it is not obvious whether they are smooth for the given coordinate of $(T_0,\infty]$ naturally arising from the standard gluing construction. This issue has recently been raised and asked to the present authors by several mathematicians.[2]

There are several ways to resolve this problem. One approach is rather topological which uses the fact that the gluing chart is smooth in the T-slice on which the gluing parameter T above is fixed. This approach is strong enough to establish all the results of [**FOn**] for which the method used in [**McSa**] was good enough to work out the analytic details along this approach. However it is not clear to the authors whether this approach is good enough to establish smoothness of the Kuranishi map or of the coordinate changes at $T=\infty$. This point was mentioned by the first and the fourth named authors themselves in [**FOn**, Remark 13.16]. To prove an existence of the Kuranishi structure that literally satisfies its axioms, we take a gluing method different from the ones employed in [**McSa**, **FOn**], called the alternating method, in this article. (See Remark 6.18 for a discussion about several other differences on the analytic points.)

Using the alternating method described in [**FOOO1**, Section A1.4], we can find an appropriate coordinate chart at $T=\infty$ so that the Kuranishi map and the coordinate changes of our Kuranishi neighborhoods are of C^∞ class. For this purpose,

[1] We can generalize it to the case of pseudoholomorphic curves attached to totally real submanifolds for given almost complex structure J on X. We do not discuss it here since we do not know its application.

[2] Among others, Y. Ruan, C. C. Liu, J. Solomon, I. Smith and H. Hofer asked the question. We thank them for asking this question.

we take the parameter $s = 1/T$. As we mentioned in [**FOOO1**, page 771] this parameter s is different from the one taken in algebraic geometry when the target X is projective. The parameter used in algebraic geometry corresponds to $z = e^{-\mathrm{const}\cdot T}$. See Chapter 8. It seems likely that in our situation where either almost complex structure is non-integrable and/or the Lagrangian submanifold enters as the boundary condition (the source being a bordered stable curve) the Kuranishi map or the coordinate changes is *not* smooth with respect to the parameter $z = e^{-\mathrm{const}\cdot T}$. But it is smooth in our parameter $s = 1/T$, as is proved in [**FOOO1**, Proposition A1.56].

The proof of this smoothness relies on some exponential decay estimate of the solution of the equation (1.1) with respect to T, that is, the length of the neck. An outline of the proof of this exponential decay is given in [**FOOO1**, Section A1.4, Lemma A1.58]. Because of the popular demand of more details of this smoothness proof, we provide such details in the present paper by polishing the presentation given in [**FOOO2**, Part 3]. The exponential decay estimate is proved as Theorem 6.4. We then show how it enables us to prove the smoothness of the coordinate change of Kuranishi structure in Chapter 8. See **Theorems 8.25 and 8.32**.

In the present paper, we restrict ourselves to the case where we glue two (bordered) stable maps whose domain curves (without considering the maps) are already stable. By restricting ourselves to this case we can address all the analytic issues needed to work out the general case also without making the notations so much heavy.

In case when the element is a stable map from a curve which has a single nodal singularity, its neighborhood still contains a stable map from a nonsingular curve. So we need to study the problem of gluing or of stretching the neck. Such a problem of gluing solutions of non-linear elliptic partial differential equation has been studied extensively in gauge theory and in symplectic geometry during the last decade of the twentieth century. Several methods had been developed to solve the problem which are also applicable to our case. In this article, following [**FOOO1**, Section A1.4], we employ the alternating method, which was first exploited by Donaldson [**D2**] in gauge theory.

REMARK 1.1. In [**D2**] Donaldson used the alternating method by inductively solving *nonlinear* equation (ASD-equation) on each of the two pieces into which he divided the given 4 manifold. In this paper we will solve the *linearized* version of Cauchy-Riemann equation inductively on each of the divided pieces. So our proof is a mixture of the alternating method and Newton's iteration.

In this method, one solves the linearization of the given equation on each piece of the domain (that is the completion of the complement of the neck region of the source of our pseudoholomorphic curve.) Then we construct a sequence of maps that converges to a solution of a version of the Cauchy-Riemann equation, that is,

$$\overline{\partial} u' \equiv 0 \mod \mathcal{E}(u') \tag{1.1}$$

and which are parameterized by a manifold (or an orbifold). Here $\mathcal{E}(u')$ is a family of finite dimensional vector spaces of smooth sections of an appropriate vector bundle depending on u'.

We provide the relevant analytic details using the same induction scheme as [**FOOO1**, Section A1.4, page 773-776]. The only difference is that we use L^2_m space (the space of maps whose derivative up to order m are of L^2 class) here, while we used L^p_1 space following the tradition of symplectic geometry community in [**FOOO1**, Section A1.4]. It is well-known that the shift of T causes loss of differentiability of the maps in terms of the order of Sobolev spaces. However by considering various *weighted* Sobolev spaces with various m simultaneously and using the definition of C^∞-topology, which is a Frechêt topology, we can still get the differentiability of C^∞ order and its exponential decay. See Remark 6.17.

In Chapter 5 we provide the details of the estimate and show that the induction scheme of [**FOOO1**, Section A1.4] provides a convergent family of solutions of our equation (1.1). This estimate is actually a fairly straightforward application of the (improved) Newton's iteration scheme although its detail is tedious to write down.

We then prove the exponential decay estimate of the derivative of the gluing map with respect to the gluing parameter in Chapter 6. See **Theorem 6.4**.

We remark that in this paper we intentionally avoid using the framework of Hilbert or Banach manifold in our gluing analysis. Certainly one can use such a framework to interpret the proof given in Chapter 5 as a kind of implicit function theorem. (Recall that Newton's iteration, which we use therein, is one of the ways to prove the implicit function theorem.) The reason we avoid using such an infinite dimensional manifold framework is because in the proof of Chapter 6 where we study T-derivatives, we need to deal with the situation where domains of the maps vary. Handling with a family of Sobolev spaces of maps with varying domain raises a nontrivial issue especially when one tries to regard the total space of the whole family as a certain version of Hilbert or Banach manifolds. Such a fact has been recognized by many researchers in various branches of mathematics. For example, the domain rotation on the Sobolev loop space is just continuous not smooth [**Kl**]. For an infinite dimensional representation, say the regular representation, of a Lie group G, the representation space (which is a space of maps from G) is not acted by the Lie algebra of G, since G action is not differentiable. This is the reason the notion of "smooth vector" is introduced. Another example is the issue of smoothness of the determinant line bundle for a smooth family of operators of Dirac type, for which the relevant family of Sobolev spaces of sections do not make a smooth Banach bundle over the parameter space (see, e.g., [**Q**], [**BF**]). It may be also worth mentioning that the method of finite dimensional approximations is used in such situations.

In the study of pseudoholomorphic curves non-smoothness of the action of the group of diffeomorphism of the domain on the Sobolev spaces is emphasized by H. Hofer, K. Wysocki and E. Zehnder [**HWZ**].

We also remark that this issue is the reason we work with L^2_m space rather than L^p_1 space. The proof of Theorem 6.4 given in this paper does *not* work if we replace L^2_m by L^p_1. See Remark 6.17.

We bypass this problem by specifying the function spaces of maps we use (which are Hilbert spaces) and explicitly defining the maps between the function spaces, e.g., the cut-paste maps entering in the gluing construction. Keeping track of the function spaces we use at all the stages of the construction, rather than putting them in an abstract framework or in a black box of functional analysis, is crucial to perform such a cut-paste process precisely in the way that the relevant derivative estimates can be obtained. It is also useful for the study of the properties of the Kuranishi structure we obtain, like its relationship to the forgetful map, to the target space group action and etc. (Note we use a Hilbert manifold

$\mathrm{Map}_{L^2_{m+1}}((K_i^S, K_i^S \cap \partial\Sigma_i), (X, L))$ in Chapter 6. There we carefully ensure that the domain and codomain of the maps do not change.)

There are several appendices which collect some parts of the proofs of Chapters 5 and 6. They comprise technical details which are based on some straightforward estimates that do not break the mainstream of the proofs. They mainly rely on the fact that for a given smooth map $F : M \to N$ the assignment $u \mapsto F(u)$ for $u : \Sigma \to M$ defines a smooth map $L^2_m(\Sigma, M) \to L^2_m(\Sigma, N)$ if m is sufficiently large. The readers who have strong background in analysis may find those appendices something obvious or at least well-known to them. We include the details of those proofs only for completeness' sake.

This paper grows up from [**FOOO2**, Part 3] (The contents of Chapter 8 is also taken from [**FOOO2**, Part 4].), which is a slightly modified version of the document which the authors uploaded to the google group Kuranishi in 2012 June. We would like to thank all the participants in the discussions of the 'Kuranishi' google group for motivating us to go through this painstaking labour. The authors also thank M. Abouzaid for making us important comments on [**FOOO2**, Part 3]. We thank the anonymous referees for going through the painstaking details of the paper, and much appreciate her/him of making a valuable assesment on the significance and usefulness of the result of the paper.

CHAPTER 2

Preliminaries

In this chapter, we collect various known results that will be used in relation to the gluing analysis, and provide basic setting for the study of gluing, and of exponential decay of the elements near the boundary of the moduli space of pseudoholomorphic maps from bordered Riemann surfaces.

2.1. Choice of Riemannian metric

We take the following choice of Riemannian metric for convenience. The proof of the next lemma can be borrowed from [**Ye**, page 683]. (See Remark A.2 for a relevant remark.)

LEMMA 2.1. *Let (X, J) be an almost complex manifold and L a (maximally) totally real submanifold with respect to J. There exists a Hermitian metric g on X such that L is totally geodesic and satisfies*

$$JT_pL \perp T_pL \tag{2.1}$$

for every $p \in L$.

PROOF. Let $\{e_1, \ldots, e_n\}$ be a local orthonormal frame on L. Since L is totally real with respect to J, $\{e_1, \ldots, e_n, Je_1, \ldots, Je_n\}$ is a local frame of TX on L. By a partition of unity we obtain a metric $\widetilde{g}$ on the vector bundle for which it satisfies $\widetilde{g}(e_i, Je_j) = 0$ for all i, j. We then extend $\widetilde{g}$ on $TX|_N$ to a tubular neighborhood of N. Then we set $\overline{g} = \widetilde{g}(J, J) + \widetilde{g}$.

Then we note that any metric on a vector bundle $E \to N$ (e.g., $E = TN$) can be extended to a metric on E as a manifold so that the latter becomes reflection invariant on E. In particular the zero section of E becomes totally geodesic. Using this, it is easy to enhance the above metric g so that L becomes totally geodesic in addition with respect to the resulting metric g. □

2.2. Exponential map and its inverse

Let ∇ be the Levi-Civita connection of the above chosen metric g in Lemma 2.1. We also use an exponential map. (The same map was used in [**FOOO1**, pages 410-411].) Denote by $\mathrm{Exp} : TX \to X$ the (global) exponential map. We put

$$\widetilde{\mathrm{Exp}}(x, v) = (x, \mathrm{Exp}(x, v)) \tag{2.2}$$

and denote by $\widetilde{\mathrm{E}} : U \to TX$ its inverse, defined on a neighborhood of the diagonal $\Delta \subset X \times X$. We define $\mathrm{E} : U \to TX$ by

$$\widetilde{\mathrm{E}}(x, y) = (x, \mathrm{E}(x, y)) \tag{2.3}$$

Note U can be defined to be

$$U = \{(x, y) \in X \times X \mid d(x, y) < \iota_X\}$$

where ι_X is the injectivity radius of (X, g). Here and hereafter $d(x, y)$ is the Riemannian distance between x and y.

REMARK 2.2. When the metric is flat and in flat coordinates, the maps introduced above also can be expressed as

$$\mathrm{Exp}(x, v) = x + v, \quad \mathrm{E}(x, y) = y - x.$$

Readers may find it useful to compare the invariant expression via the exponential maps in the present paper with these coordinate calculations used in [**FOOO2**] to get some help for visualization of exponential maps. Because of Lemma E.1, the difference between the two estimates will be exponentially small for the pseudoholomorphic maps $u(\tau, t)$ in the various circumstances we are looking at in the present paper.

2.3. Parallel transportation

We take and fix a Riemannian metric on X that satisfies the conclusion of Lemma 2.1. We denote by ι_X the injectivity radius. Namely if $x, y \in X$ with $d(x, y) < \iota_X$ then there exists a unique geodesic of length $d(x, y)$ joining x and y.

For two points $x, y \in X$ with $d(x, y) < \iota_X$ we denote by

$$\mathrm{Pal}_x^y : T_x X \to T_y X \tag{2.4}$$

the parallel transport along the unique minimal geodesic. (We use the Levi-Civita connection to define a parallel transport.) Suppose X has an almost complex structure J. We then denote by

$$(\mathrm{Pal}_x^y)^J : T_x X \to T_y X, \tag{2.5}$$

the complex linear part of Pal_x^y. Namely we decompose (uniquely) Pal_x^y to

$$\mathrm{Pal}_x^y = (\mathrm{Pal}_x^y)^J + ((\mathrm{Pal}_x^y)^J)'$$

such that

$$(\mathrm{Pal}_x^y)^J \circ J_x = J_y \circ (\mathrm{Pal}_x^y)^J, \quad ((\mathrm{Pal}_x^y)^J)' \circ J_x = -J_y \circ ((\mathrm{Pal}_x^y)^J)'$$

In other words

$$(\mathrm{Pal}_x^y)^J = \frac{1}{2}\Big(\mathrm{Pal}_x^y - J_y \circ \mathrm{Pal}_x^y \circ J_x\Big).$$

We choose and fix a constant $\iota'_X > 0$ such that the following holds.

CONDITION 2.3. If $d(x, y) \le \iota'_X$ then $(\mathrm{Pal}_x^y)^J$ is a linear isomorphism.

Let Σ be a two dimensional complex manifold and $u, v : \Sigma \to X$. We assume

$$\sup\{d(u(z), v(z)) \mid z \in \Sigma\} \le \iota'_X.$$

Then using pointwise maps Pal_x^y and $(\mathrm{Pal}_x^y)^J$ we obtain the maps of sections

$$\mathrm{Pal}_u^v : \Gamma(\Sigma, u^*TX) \to \Gamma(\Sigma, v^*TX), (\mathrm{Pal}_u^v)^J : \Gamma(\Sigma, u^*TX) \to \Gamma(\Sigma, v^*TX). \tag{2.6}$$

We also note that by composing with the $(0, 1)$-projections with respect to $J : TM \to TM$, $(\mathrm{Pal}_x^y)^J$ also induces the map

$$(\mathrm{Pal}_{u(x)}^{v(x)})^{(0,1)} : T_{u(x)}X \otimes \Lambda_x^{0,1} \to T_{v(x)}X \otimes \Lambda_x^{0,1}$$

which in turn induces the map of sections

$$(\mathrm{Pal}_u^v)^{(0,1)} : \Gamma(\Sigma, u^*TX \otimes \Lambda^{0,1}) \to \Gamma(\Sigma, v^*TX \otimes \Lambda^{0,1}). \tag{2.7}$$

2.4. Exponential decay of pseudoholomorphic maps with small image

We now consider a pseudoholomorphic map $u : [-S, S] \times [0,1] \to (X, L)$

$$\frac{\partial u}{\partial \tau} + J\frac{\partial u}{\partial t} = 0$$

with finite energy $\mathfrak{E}(u) < \infty$. We also recall the definition of energy

$$\mathfrak{E}(u) = \frac{1}{2}\int \left(\left|\frac{\partial u}{\partial \tau}\right|^2 + \left|\frac{\partial u}{\partial t}\right|^2 \right) dt\, d\tau$$

where the norm $|\cdot|$ is measured in terms of a metric g that is compatible with ω, $g = \omega(\cdot, J\cdot)$ for a fixed compatible almost complex structure J. We use the following well-known uniform exponential decay estimate for pseudoholomorphic curve with small diameter.

LEMMA 2.4. *There exists $T_{(2.8)} > 0$ and $\delta_1 > 0, \epsilon_1 > 0, C_{(2.8)} > 0$ depending only on X, L, and $\mathfrak{E}_0$ with the following properties. If $T > T_{(2.8)}$, $u : [-T-1, T+1] \times [0,1] \to X$ is pseudoholomorphic, $\mathfrak{E}(u) \le \mathfrak{E}_0$, $u([-T-1, T+1] \times \{0,1\}) \subset L$ and* Diam(Im*u*) $\le \epsilon_1$*, then*

$$\left|\frac{\partial u}{\partial \tau}(\tau, t)\right| + \left|\frac{\partial u}{\partial t}(\tau, t)\right| \le C_{(2.8)} e^{-\delta_1 d(\tau, \partial[-T-1, T+1])} \tag{2.8}$$

for $(\tau, t) \in [-T, T] \times [0,1]$.

We refer to [**FOn**, Lemma 11.2] or to [**Oh2**, Lemma B.1] for its proof, for example.

Using this C^1-exponential decay, ϵ-regularity theorem and the uniform local a priori estimates, we also obtain

LEMMA 2.5. *There exists $C_{k,(2.9)}$ depending only on the $(X, J, g), L$, $\mathfrak{E}_0$ and $k \ge 0$ such that if $u : [-T-1, T+1] \times [0,1] \to X$ is as in Lemma 2.4 then*

$$\left|\frac{\partial u}{\partial \tau}(\tau, t)\right|_{C^k} \le C_{k,(2.9)} e^{-\delta_1 d(\tau, \partial[-T-1, T+1])}. \tag{2.9}$$

In particular, if $u : [0, \infty) \times [0,1] \to X$ (or $u : (-\infty, 0] \times [0,1] \to X$) is pseudoholomorphic $u([0,\infty) \times \{0,1\}) \subset L$ (or $u((-\infty, 0] \times \{0,1\}) \subset L$), Diam(Im*u*) $\le \epsilon_1$ *and $\mathfrak{E}(u) \le \mathfrak{E}_0$, and $k \ge 0$ then we have*

$$\left|\frac{\partial u}{\partial \tau}(\tau, t)\right|_{C^k} \le C_{k,(2.9)} e^{-\delta_1 |\tau|}. \tag{2.10}$$

REMARK 2.6. Actually we can take $\delta_1 = \pi$. We do not need this explicit value in this paper.

CONVENTION 2.7. Hereafter we assume that *all* pseudoholomorphic curves u we consider satisfy the inequality

$$\mathfrak{E}(u) \le \mathfrak{E}_0,$$

for some fixed $\mathfrak{E}_0$. So the $\mathfrak{E}_0$ dependence of various constants will not be mentioned.

REMARK 2.8. Since many constants appear in this paper we use the following enumeration convention. The constant such as, $C_{m,(10.12)}$, for example, is the one appearing in formula (10.12) which may depend on m.

For some constants appearing repeatedly, we do *not* apply this rule of enumerating by the number of the associated formulae. The list of such constants is: ι'_X appears in Condition 2.3, ϵ_1, δ_1 appear in Lemma 2.4, δ appears in (3.11), $T_{1,\epsilon(1)}$, $T_{2,\epsilon(1),\epsilon(2)}$ appear in Theorem 3.13, $C_{1,m}$ appear Lemma 5.2, ϵ_2 appears in Lemma 5.6, $C_{2,m}$ appears in Lemma 5.15, $C_{3,m}$ appears in Lemma 5.17, $C_{4,m}$ appears in (5.69), $\epsilon_{3,m}$ and $T_{3,m,\epsilon(4)}$, $T_{4,m}$ appear in Proposition 5.23, $C_{5,m}$, $C_{6,m}$, $C_{7,m}$, $C_{8,m}$, $C_{9,m}$ $T_{5,m,\epsilon(6)}$, $T_{6,m}$, $\epsilon_{4,m}$ appear in Proposition 6.9, ϵ_5 appears in Proposition 7.1. δ_2 appears in Theorem 8.17, δ_3 appears in Proposition 8.19 and δ_4 appears in Proposition 8.35, ϵ_6, ϵ_7, ν_1, ν_2 appear around (8.45), ϵ_8, ϵ_9, ϵ_{10}, $\nu_{3,m}$, $\nu_{4,m}$ appear in Lemma 8.24.

The positive numbers $\epsilon(1), \epsilon(2), \ldots$ are *not* fixed but may vary.

CHAPTER 3

Statement of the gluing theorem

For simplicity and clarity of the exposition on the main analytical points, we consider the case where we glue pseudoholomorphic maps from two stable bordered Riemann surfaces to (X, L) in the present paper. The general case is proved in the same way in [**FOOO2**, Part IV].

3.1. The weighted Sobolev spaces for the gluing analysis

Let Σ_i be a bordered Riemann surface for $i = 1, 2$. We assume that each Σ_i has strip-like ends such that $\Sigma_i \setminus K_i$ becomes semi-infinite strip. Recall that Σ_i is conformally a bordered Riemann surface with one puncture.

More specifically, we fix a Kähler metric on Σ_i whose end is isometric to the flat strip. We fix isothermal coordinates (τ, t) so that Σ_i is decomposed into

$$\begin{aligned} \Sigma_1 &= K_1 \cup ([-5T, \infty) \times [0,1]), \\ \Sigma_2 &= ((-\infty, 5T] \times [0,1]) \cup K_2 \end{aligned} \tag{3.1}$$

where K_i is a compact subset of Σ_i respectively. See Figure 3.1.

Here (τ, t) is the standard coordinates on the strip $\mathbb{R} \times [0,1]$ and restricted to $[-5T, \infty) \times [0,1]$ (resp. $(-\infty, 5T] \times [0,1]$) on $\Sigma_1 \setminus K_1$ (resp. $\Sigma_2 \setminus K_2$). We would like to note that the coordinate τ depends on T. For each $T > 0$, we put

$$\Sigma_T = K_1 \cup ([-5T, 5T] \times [0,1]) \cup K_2. \tag{3.2}$$

See Figure 3.2.

Let X be a compact symplectic manifold with compatible almost complex structure and L its Lagrangian submanifold.

Let $u_i : (\Sigma_i, \partial \Sigma_i) \to (X, L)$, $i = 1, 2$ be pseudoholomorphic maps of finite energy. Then, by the removable singularity theorem [**Oh1**], the maps u_i smoothly

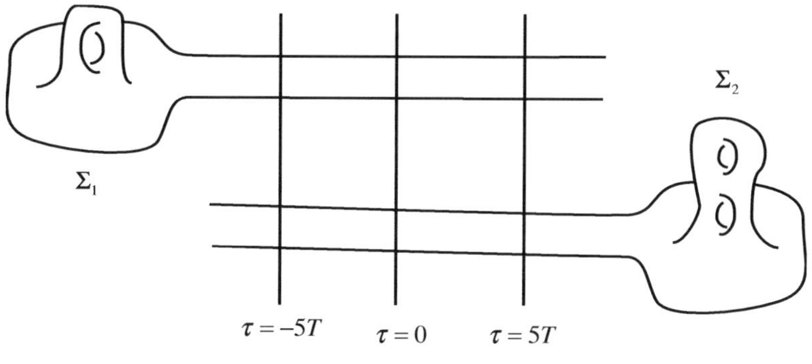

FIGURE 3.1. Σ_1 and Σ_2.

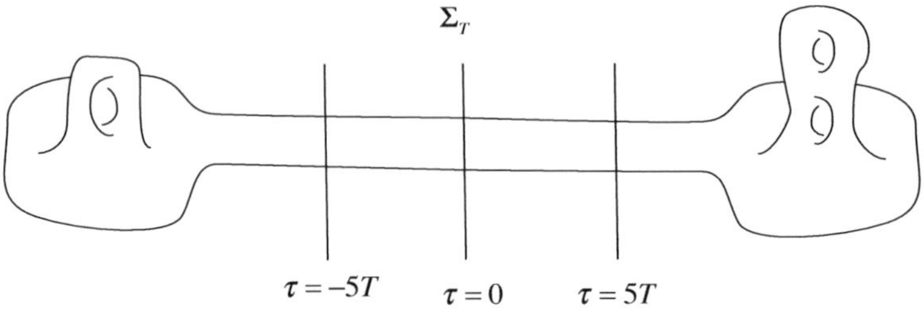

FIGURE 3.2. Σ_T.

extend to the associated puncture of Σ_i and so the asymptotic limits

$$\lim_{\tau \to \infty} u_1(\tau, t) \in L \tag{3.3}$$

and

$$\lim_{\tau \to -\infty} u_2(\tau, t) \in L \tag{3.4}$$

are uniquely determined and do not depend on $t \in [0, 1]$.

We will consider a pair (u_1, u_2) of the maps for which the limits (3.3), (3.4) coincide which we denote by $p_0 \in L$.

CONDITION 3.1. We also assume that

$$\text{Diam } u_i(\Sigma_i \setminus K_i) \le \epsilon_1 \tag{3.5}$$

where ϵ_1 is as in Lemma 2.4.

Then the exponential decay (2.10) holds by Lemma 2.5.

REMARK 3.2. If u_1 (resp. u_2) satisfies (3.3) (resp. (3.4)) then we may replace K_1 (resp. K_2) by $K_1 \cup [-5T, -5T + S] \times [0, 1]$ (resp. $K_2 \cup [-S + 5T, 5T] \times [0, 1]$) and change the T coordinate to $T - S$ (resp. $T + S$) for an appropriate S so that (3.5) is satisfied. In other words, we can assume (3.5) without loss of generality.

For the clarity of our exposition, we consider the case of fixed complex structures of the domains Σ_i, in Chapters 3-7. We then include the deformation parameter of the complex structures of Σ_i in Chapter 8. In order to handle the case of genus 0, we also need to stabilize and so add marked points on the domain curve Σ_i. We denote the ordered set of marked points by $\vec{z}_i = (z_{i,1}, \cdots, z_{i,k_i})$ for $i = 1, 2$. For the simplicity of notations, we only consider boundary marked points.

We may add one point $\tau = \infty$ or $-\infty$ and compactify the source curve Σ_i and extend our map u_i thereto. We regard the added point as the 0-th marked point $z_{i,0}$.

ASSUMPTION 3.3. In this paper we always assume that $(\Sigma_i, \vec{z}_i)$ is stable.

Note there is one more marked point at infinity other than $\vec{z}_i$. We also remark that here Σ_i is assumed to have no interior or boundary node. So if Σ_i is not a disc then it is stable and if Σ_i is a disc, the condition $k_i \ge 2$ is equivalent to the stability.

We will define a map:

$$D_{u_i}\overline{\partial} : W^2_{m+1,\delta}((\Sigma_i, \partial\Sigma_i); u_i^*TX, u_i^*TL) \to L^2_{m,\delta}(\Sigma_i; u_i^*TX \otimes \Lambda^{0,1}), \tag{3.6}$$

which is the linearization of $\overline{\partial}$ at u_i. Here we denote $\overline{\partial} = \overline{\partial}_J$ by omitting J. The domain, that is, the function space

$$W^2_{m+1,\delta}((\Sigma_i, \partial\Sigma_i); u_i^*TX, u_i^*TL),$$

is defined below.

We assume that the constant ϵ_1 given in Condition 3.1 is at least smaller than ι'_X given in Condition 2.3. Then to each $v \in T_{p_0}L$, we define the section $v^{\rm Pal} \in \Gamma([-5T, 5T] \times [0,1]; u_1^*TX)$ by

$$v^{\rm Pal}(\tau, t) \equiv {\rm Pal}_{p_0}^{u_i(\tau,t)}(v). \tag{3.7}$$

We note that the sections (3.7) satisfy the boundary condition at $t = 0, 1$ because we use the Levi-Civita connection of a metric with respect to which the Lagrangian submanifold L is totally geodesic. (See Lemma 2.1.)

More generally if $x \in X$, $v \in T_xX$ and $u : \Sigma_0 \to X$ is a map to the ι_X neighborhood of x from a 2 dimensional manifold, we define $v^{\rm Pal} \in \Gamma(\Sigma_0, u^*TX)$ by

$$v^{\rm Pal}(z) = {\rm Pal}_x^{u(z)}(v). \tag{3.8}$$

DEFINITION 3.4 ([**FOOO1**, Section 7.1.3]). [1] Let $p_0 \in L$ be the point given in (3.3) or in (3.4). Denote by $L^2_{m+1,loc}((\Sigma_i, \partial\Sigma_i); u_i^*TX; u_i^*TL)$ the set of the sections s of u_i^*TX which are locally of L^2_{m+1}-class. (Namely its differential up to order $m+1$ is locally of L^2-class. Here m is sufficiently large, say larger than 10.) We also assume $s(z) \in u_i^*TL$ for $z \in \partial\Sigma_i$. We define a Hilbert space

$$W^2_{m+1,\delta}((\Sigma_i, \partial\Sigma_i); u_i^*TX, u_i^*TL)$$

as the set of all pairs (s, v) where $s \in L^2_{m+1,loc}((\Sigma_i, \partial\Sigma_i); u_i^*TX; u_i^*TL)$, $v \in T_{p_0}L$ such that

$$\sum_{k=0}^{m+1} \int_{\Sigma_i \setminus K_i} e^{2\delta|\tau \pm 5T|} |\nabla^k (s - v^{\rm Pal})|^2 < \infty. \tag{3.9}$$

Here $\pm = +$ for $i = 1$ and $\pm = -$ for $i = 2$. We define its norm $\|(s,v)\|_{W^2_{m+1,\delta}(\Sigma_i)}$ by

$$\|(s,v)\|^2_{W^2_{m+1,\delta}(\Sigma_i)} = (3.9) + \sum_{i=1,2} \sum_{k=0}^{m+1} \int_{K_i} |\nabla^k s|^2 + \|v\|^2. \tag{3.10}$$

Hereafter we omit Σ_i from the notation $\|(s,v)\|_{W^2_{m+1,\delta}(\Sigma_i)}$ in case the domain is obvious from the context. We denote by $L^2_{m+1,\delta}((\Sigma_i, \partial\Sigma_i); u_i^*TX, u_i^*TL)$ the subspace of $W^2_{m+1,\delta}((\Sigma_i, \partial\Sigma_i); u_i^*TX, u_i^*TL)$ consisting of elements with $v = 0$.

The space $L^2_{m,\delta}(\Sigma_i; u_i^*TX \otimes \Lambda^{0,1})$ is defined similarly without boundary condition and without v. (See (4.10).)

Since the relevant s for $T = \infty$ does not decay to 0 but converges to an element $v \in T_{p_0}X$, we denote the subscript of $\|s\|^2_{W^2_{m+1,\delta}}$ to be $W^2_{m+1,\delta}$ instead of $L^2_{m+1,\delta}$.

We let $D_{u_i}\overline{\partial}$ act trivially on the v-component and just act on the s-component of $(s, v) \in W^2_{m+1,\delta}((\Sigma_i, \partial\Sigma_i); u_i^*TX, u_i^*TL)$. We choose $\delta > 0$ with

$$\delta \le \frac{\delta_1}{10} \tag{3.11}$$

[1] In [**FOOO1**] L^p_1 space is used in stead of L^2_m space.

where $\delta_1 > 0$ is the constant given in Lemma 2.4. (We refer (6.38), (6.74) and the discussion therearound for the reason we make such a choice.) *We never change this constant δ in the rest of this paper.* Lemma 2.5 (2.10) implies

$$\big|(D_{u_i}\overline{\partial})(v^{\mathrm{Pal}})\big|_{C^k} < C_{k,(3.12)}C_{k,(2.9)}e^{-\delta_1|\tau|}\|v\|. \tag{3.12}$$

Therefore the operator (3.6), especially with $W^2_{m+1,\delta}((\Sigma_i, \partial\Sigma_i); u_i^*TX, u_i^*TL)$ as its domain, is defined and bounded. It is a standard fact (see [**LM**]) that this operator is also Fredholm if we choose δ sufficiently small. In the current context, δ depends only on the geometry of (X, g, J) and $L \subset X$.

REMARK 3.5. We would like to note that $W^2_{m+1,\delta}((\Sigma_i, \partial\Sigma_i); u_i^*TX, u_i^*TL)$ is a completion of the set of infinitesimal variations of $u_i : \Sigma_i \to X$ satisfying the boundary conditions

$$u_i(\partial\Sigma_i) \subset L, \qquad \lim_{\tau\to+\infty} u_1(\tau, t) = \lim_{\tau\to-\infty} u_2(\tau, t) = p_0 \in L$$

in particular allowing the point p_0 to vary inside L. We have an exact sequence

$$\begin{aligned} 0 &\to L^2_{m+1,\delta}((\Sigma_i, \partial\Sigma_i); u_i^*TX, u_i^*TL) \\ &\to W^2_{m+1,\delta}((\Sigma_i, \partial\Sigma_i); u_i^*TX, u_i^*TL) \to T_{p_0}L \to 0 \end{aligned}$$

where the first map is the inclusion $s \mapsto (s, 0)$ and the second map is the asymptotic evaluation of the section s at $\pm\infty$ for $i = 1$, 2 respectively.

We next define our (family of) obstruction spaces $\mathcal{E}_i(u')$.

We fix a pair of pseudoholomorphic maps $u_i^{\mathfrak{ob}} : (\Sigma_i, \partial\Sigma_i) \to (X, L)$ for $i = 1, 2$. We assume

$$d(u_i(z), u_i^{\mathfrak{ob}}(z)) \le \frac{\iota'_X}{2} \tag{3.13}$$

where ι'_X is as in (2.3).

We take a finite dimensional linear subspace

$$\mathcal{E}_i^{\mathfrak{ob}} = \mathcal{E}_i(u_i^{\mathfrak{ob}}) \subset \Gamma(K_i; (u_i^{\mathfrak{ob}})^*TX \otimes \Lambda^{0,1}), \quad i = 1,\ 2 \tag{3.14}$$

which consists of smooth sections of $(u_i^{\mathfrak{ob}})^*TX \otimes \Lambda^{0,1}$ supported in a compact subset K_i^0 of Int K_i.

REMARK 3.6. When we use Theorems 3.13 and 6.4 for the construction of the Kuranishi structure on a moduli space of bordered pseudoholomorphic curves, we take several $u_i^{\mathfrak{ob}}$'s and associate $\mathcal{E}_i^{\mathfrak{ob}}$ to each of them. Then in place of $\mathcal{E}_1(u') \oplus \mathcal{E}_2(u')$ we take a sum of finitely many of them. See [**FOn**, page 1003] and [**FOOO2**, Definition 18.12]. For the simplicity of exposition, we focus on the case when we have one set of $u_i^{\mathfrak{ob}}$, $\mathcal{E}_i^{\mathfrak{ob}}$ in this paper. The case of several such $u_i^{\mathfrak{ob}}$'s and $\mathcal{E}_i^{\mathfrak{ob}}$'s is treated in the same way as that of Theorems 3.13 and 6.4.

3.2. Analytic framework for gluing

We state the first half of the main result as Theorem 3.13 below. The second half is Theorem 6.4. Actually Theorem 3.13 itself is a classical result. We provide its statement here because we write its proof in such a way that is suitable for the proof of derivative estimates given in Theorem 6.4. Let

$$u' : (\Sigma_T, \partial\Sigma_T) \to (X, L) \tag{3.15}$$

be a smooth map.

We consider the following conditions for given constant $\epsilon > 0$.

CONDITION 3.7.
(1) $u'|_{K_i}$ is ϵ-close to $u_i|_{K_i}$ in the C^1 sense. [2]
(2) The diameter of $u'(X \setminus (K_1 \cup K_2))$ is smaller than ϵ.

We define the map

$$(\text{Pal}_{u_i^{\mathfrak{ob}}}^{u'})^{(0,1)} : \Gamma(K_i; (u_i^{\mathfrak{ob}})^*TX \otimes \Lambda^{0,1}) \to \Gamma(\Sigma_T; (u')^*TX \otimes \Lambda^{0,1}) \tag{3.16}$$

by composing the map $(\text{Pal}_{u_i^{\mathfrak{ob}}}^{u'})^{(0,1)} : \Gamma(K_i; (u_i^{\mathfrak{ob}})^*TX \otimes \Lambda^{0,1}) \to \Gamma(K_i; (u')^*TX \otimes \Lambda^{0,1})$ in (2.7) with the map induced by the inclusion $K_i \subset \Sigma_T$.

If we take $\epsilon > 0$ sufficiently small, then (3.13) and Condition 3.7 (1) imply that this map (3.16) is defined for such u'.

We then define the map

$$I_{u',i} : \mathcal{E}_i^{\mathfrak{ob}} \to \Gamma(\Sigma_T; (u')^*TX \otimes \Lambda^{0,1}) \tag{3.17}$$

to be the restriction of $(\text{Pal}_{u_i^{\mathfrak{ob}}}^{u'})^{(0,1)}$ to $\mathcal{E}_i^{\mathfrak{ob}}$ and put

$$\mathcal{E}_i(u') = I_{u',i}(\mathcal{E}_i^{\mathfrak{ob}}). \tag{3.18}$$

Recall that elements of $\mathcal{E}_i^{\mathfrak{ob}}$ are supported on $K_i^0 \subset \text{Int}\, K_i$. Therefore for this definition, we first take the parallel transport[3] of $\eta \in \mathcal{E}_i^{\mathfrak{ob}}$ along the shortest geodesic between $u_i^{\mathfrak{ob}}(z)$ and $u'(z)$ for $z \in K_i^0 \subset \Sigma_T$ and then extend the resulting section by zero to the rest of Σ_T.

The equation we study in the rest of the paper is of the form

$$\overline{\partial} u' \equiv 0, \qquad \text{mod } \mathcal{E}_1(u') \oplus \mathcal{E}_2(u'). \tag{3.19}$$

DEFINITION 3.8. We denote by $\mathcal{M}^{\mathcal{E}_1 \oplus \mathcal{E}_2}((\Sigma_T, \vec{z}); u_1, u_2)_\epsilon$ the set of solutions of (3.19) satisfying the Condition 3.7.

Theorem 3.13 will describe all the solutions of (3.19) 'sufficiently close to the (pre-)glued map $u_1 \# u_2$.' To make this statement precise, we need to prepare more notations.

Let $u_i' : (\Sigma_i, \partial \Sigma_i) \to (X, L)$ be any smooth map, not necessarily pseudoholomorphic. We put the following conditions on u_i'.

CONDITION 3.9.
(1) $u_i'|_{K_i}$ is ϵ-close to $u_i|_{K_i}$ in C^1 sense.
(2) The diameter of $u_1'(\Sigma_1 \setminus K_1)$, (resp. $u_2'(\Sigma_2 \setminus K_2)$) is smaller than ϵ.

Then we define

$$I_{u_i'} : \mathcal{E}_i^{\mathfrak{ob}} \to \Gamma(\Sigma_i; (u_i')^*TX \otimes \Lambda^{0,1}) \tag{3.20}$$

in the same way as $I_{u'}$ i.e.,

$$I_{u_i'}(\eta) := (\text{Pal}_{u_i^{\mathfrak{ob}}}^{u_i'})^{(0,1)}(\eta), \quad \eta \in \mathcal{E}_i^{\mathfrak{ob}}.$$

(This makes sense if u_i' satisfies Condition 3.9 for $\epsilon < \iota_X'/2$. In fact, then we have $d(u_i'(x), u_i^{\mathfrak{ob}}(x)) < \iota_X'$.) We put

$$\mathcal{E}_i(u_i') = I_{u_i'}(\mathcal{E}_i^{\mathfrak{ob}}). \tag{3.21}$$

[2]We use a Riemannian metric on K_i and X to define the C^1 norm we use.

[3]more precisely its complex linear part. We abuse the notation sometimes in this way.

So we can define the equation

$$\overline{\partial} u_i' \equiv 0, \qquad \text{mod } \mathcal{E}_i(u_i'). \tag{3.22}$$

We write $\mathcal{E}_i$ in place of $\mathcal{E}_i(u_i')$ sometimes.

REMARK 3.10. Note that in the current situation where all the irreducible components of Σ_i intersect with boundary and Σ_i has nontrivial boundary with at least one boundary marked point (that is, either $\tau = \infty$ or $-\infty$ depending on $i = 1$ or $i = 2$), Assumption 3.3 implies that Σ_i carries no nontrivial automorphism. In the case when there is an irreducible component which does not intersect boundary, the automorphism group can be nontrivial. We can use the standard trick to add marked points and then use codimension 2 submanifold to kill the extra freedom of moving added marked points. (See [**FOn**, Appendix] or [**FOOO2**, Part IV], for example.)

DEFINITION 3.11. We denote by $\mathcal{M}^{\mathcal{E}_i}((\Sigma_i, \vec{z}_i); u_i)_\epsilon$ the set of all maps u_i' : $(\Sigma_i, \partial\Sigma_i) \to (X, L)$ whose domain is the marked bordered Riemann surface $(\Sigma_i, \vec{z}_i)$ given as in (3.1) for $i = 1$, 2 such that

(1) u_i' satisfies (3.22).
(2) Condition 3.9 is satisfied.

When $\beta_i = [u_i] \in H_2(X, L)$ we write $\mathcal{M}^{\mathcal{E}_i}((\Sigma_i, \vec{z}_i); \beta_i)_\epsilon$ sometimes in place of $\mathcal{M}^{\mathcal{E}_i}((\Sigma_i, \vec{z}_i); u_i)_\epsilon$.

We work under the following assumption. This assumption is put on the pair (u_1, u_2) which we glue.

ASSUMPTION 3.12. Let $u_1(\infty) = u_2(-\infty) = p_0 \in L$ and let $(D_{u_i}\overline{\partial})^{-1}(\mathcal{E}_i(u_i))$ be the kernel of (3.23). We assume:

1. (Mapping transversality)

$$D_{u_i}\overline{\partial} : W^2_{m+1,\delta}((\Sigma_i, \partial\Sigma_i); u_i^*TX, u_i^*TL) \to L^2_{m,\delta}(\Sigma_i; u_i^*TX \otimes \Lambda^{0,1})/\mathcal{E}_i(u_i) \tag{3.23}$$

is surjective.

2. (Evaluation transversality) For each $i = 1$, 2, define the linearized evaluation map

$$\mathrm{Dev}_{i,\infty} : W^2_{m+1,\delta}((\Sigma_i, \partial\Sigma_i); u_i^*TX, u_i^*TL) \to T_{p_0}L \tag{3.24}$$

by

$$\mathrm{Dev}_{i,\infty}(s, v) = v.$$

Then

$$\mathrm{Dev}_{1,\infty} - \mathrm{Dev}_{2,\infty} : (D_{u_1}\overline{\partial})^{-1}(\mathcal{E}_1(u_1)) \oplus (D_{u_2}\overline{\partial})^{-1}(\mathcal{E}_2(u_2)) \to T_{p_0}L \tag{3.25}$$

is surjective.

The surjectivity of (3.24), (3.25) and the implicit function theorem imply that, if ϵ is small enough, there exists a finite dimensional vector space $\widetilde{V}_i$ and a neighborhood $V_i(\epsilon)$ of 0 therein, which provides a local parametrization of $\mathcal{M}^{\mathcal{E}_i}((\Sigma_i, \vec{z}_i); u_i)$,

$$\mathcal{M}^{\mathcal{E}_i}((\Sigma_i, \vec{z}_i); u_i)_\epsilon \cong V_i(\epsilon) \tag{3.26}$$

near the given solution u_i. For any $\rho_i \in V_i(\epsilon)$, we denote by $u_i^{\rho_i} : (\Sigma_i, \partial\Sigma_i) \to (X, L)$ the corresponding solution of (3.22).

We have an asymptotic evaluation map

$$\mathrm{ev}_{i,\infty} : \mathcal{M}^{\mathcal{E}_i}((\Sigma_i, \vec{z}_i); u_i)_\epsilon \to L.$$

Namely, we define

$$\mathrm{ev}_{i,\infty}(u'_i) = \lim_{\tau \to \pm\infty} u'_i(\tau, t).$$

(Here $\pm = +$ for $i = 1$ and $-$ for $i = 2$.).

We consider the fiber product:

$$V_1(\epsilon) \times_L V_2(\epsilon) = \mathcal{M}^{\mathcal{E}_1}((\Sigma_1, \vec{z}_1); u_1)_\epsilon \times_L \mathcal{M}^{\mathcal{E}_2}((\Sigma_2, \vec{z}_2); u_2)_\epsilon. \tag{3.27}$$

The surjectivity of (3.25) implies that the fiber product (3.27) is transversal and so $V_1(\epsilon) \times_L V_2(\epsilon)$ is a smooth manifold. We write elements of $V_1(\epsilon) \times_L V_2(\epsilon)$ as $\rho = (\rho_1, \rho_2)$.

THEOREM 3.13. *For each* $\epsilon(1) > 0$, *there exist* $\epsilon(2) > 0$, $T_{1,\epsilon(1)} > 0$ *and a map*

$$\mathrm{Glue}_T \colon \mathcal{M}^{\mathcal{E}_1}((\Sigma_1, \vec{z}_1); u_1)_{\epsilon(2)} \times_L \mathcal{M}^{\mathcal{E}_2}((\Sigma_2, \vec{z}_2); u_2)_{\epsilon(2)} \to \mathcal{M}^{\mathcal{E}_1 \oplus \mathcal{E}_2}((\Sigma_T, \vec{z}); u_1, u_2)_{\epsilon(1)}$$

for $T > T_{1,\epsilon(1)}$ *that satisfies the following. Here* $\vec{z} = \vec{z}_1 \cup \vec{z}_2$.

(1) *The map* Glue_T *is a diffeomorphism onto its image.*
(2) *There exist* $\epsilon_{\epsilon(1),\epsilon(2),\langle\langle 3.13\rangle\rangle} > 0$ *and* $T_{2,\epsilon(1),\epsilon(2)} > 0$ *depending on* $\epsilon(1)$, $\epsilon(2)$ *such that the image of* Glue_T *contains*

$$\mathcal{M}^{\mathcal{E}_1 \oplus \mathcal{E}_2}((\Sigma_T, \vec{z}); u_1, u_2)_{\epsilon_{\epsilon(1),\epsilon(2),\langle\langle 3.13\rangle\rangle}}$$

if $T > T_{2,\epsilon(1),\epsilon(2)}$.

The main result about exponential decay estimate of this map will be given in Chapter 6. (Theorem 6.4.)

CHAPTER 4

Proof of the gluing theorem I: Cut-off functions and weighted Sobolev norm

The proof of Theorem 3.13 was given in [**FOOO1**, Section 7.1.3]. Here we follow the scheme of the proof of [**FOOO1**, Section A1.4] and provide more of its details. As mentioned there, the scheme of the proof is originated from Donaldson's paper [**D2**], and its Morse-Bott version in [**Fu**].

We first introduce certain cut-off functions. First let $\mathcal{A}_T \subset \Sigma_T$ and $\mathcal{B}_T \subset \Sigma_T$ be the domains defined by , ,

$$\mathcal{A}_T = [-T-1, -T+1] \times [0,1], \qquad \mathcal{B}_T = [T-1, T+1] \times [0,1].$$

We may regard $\mathcal{A}_T, \mathcal{B}_T$ as subsets of Σ_i. The third domain is

$$\mathcal{X} = [-1,1] \times [0,1] \subset \Sigma_T.$$

See Figure 4.1. We may also regard $\mathcal{X} \subset \Sigma_i$. We recall readers the definitions of Σ_i from (3.1).

Let $\chi_{\mathcal{A}}^{\leftarrow}$, $\chi_{\mathcal{A}}^{\rightarrow}$ be the smooth functions on $[-5T, 5T] \times [0,1]$ defined by

$$\chi_{\mathcal{A}}^{\leftarrow}(\tau,t) = \begin{cases} 1 & \tau < -T-1 \\ 0 & \tau > -T+1. \end{cases} \tag{4.1}$$
$$\chi_{\mathcal{A}}^{\rightarrow} = 1 - \chi_{\mathcal{A}}^{\leftarrow}.$$

See Figure 4.2. We also define

$$\chi_{\mathcal{B}}^{\leftarrow}(\tau,t) = \begin{cases} 1 & \tau < T-1 \\ 0 & \tau > T+1. \end{cases} \tag{4.2}$$
$$\chi_{\mathcal{B}}^{\rightarrow} = 1 - \chi_{\mathcal{B}}^{\leftarrow}$$

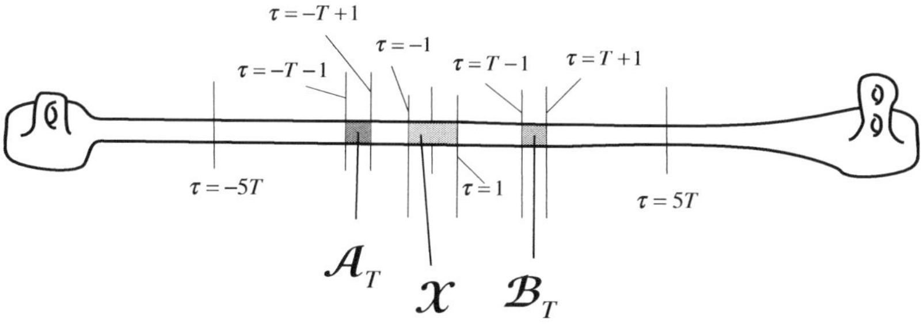

FIGURE 4.1. $\mathcal{A}_T, \mathcal{B}_T$ and $\mathcal{X}$.

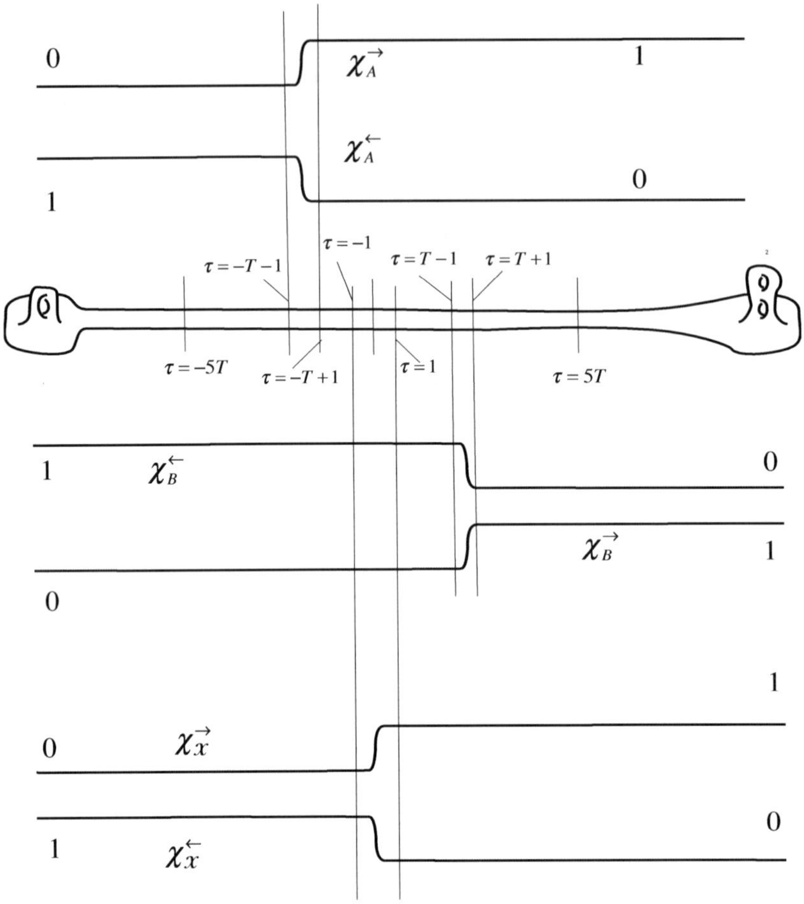

FIGURE 4.2. Cut off functions.

and

$$\chi_{\mathcal{X}}^{\leftarrow}(\tau,t) = \begin{cases} 1 & \tau < -1 \\ 0 & \tau > 1. \end{cases} \tag{4.3}$$
$$\chi_{\mathcal{X}}^{\rightarrow} = 1 - \chi_{\mathcal{X}}^{\leftarrow}.$$

We extend these functions to Σ_T and Σ_i $(i = 1, 2)$ so that they are locally constant outside $[-5T, 5T] \times [0, 1]$. We denote them by the same symbol.

We now introduce weighted Sobolev norms and their local versions for sections on Σ_T or Σ_i. We first define weight functions $e_{i,\delta} : \Sigma_i \to [1, \infty)$, $i = 1, 2$ of

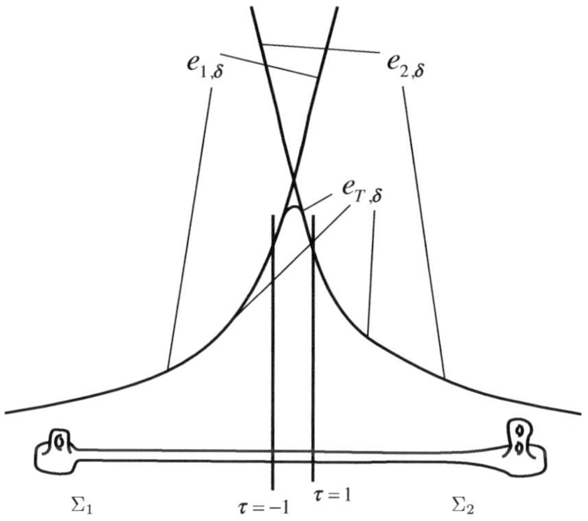

FIGURE 4.3. Weight function.

C^∞-class as follows:

$$e_{1,\delta}(\tau,t)\begin{cases}= e^{\delta(\tau+5T)} & \text{if } \tau > 1-5T\\ =1 & \text{on } K_1\\ \in [1,10] & \text{if } \tau < 1-5T\end{cases}\tag{4.4}$$

$$e_{2,\delta}(\tau,t)\begin{cases}= e^{\delta(-\tau+5T)} & \text{if } \tau < 5T-1\\ =1 & \text{on } K_2\\ \in [1,10] & \text{if } \tau > 5T-1.\end{cases}\tag{4.5}$$

(We choose $\delta > 0$ so small that it satisfies $1 < e^\delta < 10$ above.) We also define $e_{T,\delta} : \Sigma_T \to [1,\infty)$ as follows. (See Figure 4.3.)

$$e_{T,\delta}(\tau,t)\begin{cases}= e^{\delta(-\tau+5T)} & \text{if } 1 < \tau < 5T-1\\ = e^{\delta(\tau+5T)} & \text{if } -1 > \tau > 1-5T\\ =1 & \text{on } K_1\cup K_2\\ \in [1,10] & \text{if } |\tau-5T|<1 \text{ or } |\tau+5T|<1\\ \in [e^{5T\delta}/10, e^{5T\delta}] & \text{if } |\tau|<1.\end{cases}\tag{4.6}$$

The weighted Sobolev norm we use for $L^2_{m,\delta}(\Sigma_i; u_i^*TX \otimes \Lambda^{0,1})$ is given by

$$\|s\|^2_{L^2_{m,\delta}} = \sum_{k=0}^{m} \int_{\Sigma_i} e^2_{i,\delta} |\nabla^k s|^2 \mathrm{vol}_{\Sigma_i}.\tag{4.7}$$

For $(s,v) \in W^2_{m+1,\delta}((\Sigma_i,\partial\Sigma_i); u_i^*TX, u_i^*TL)$ we defined in (3.10)

$$\begin{aligned}\|(s,v)\|^2_{W^2_{m+1,\delta}} &= \sum_{k=0}^{m+1} \int_{K_i} |\nabla^k s|^2 \mathrm{vol}_{\Sigma_i}\\ &+ \sum_{k=0}^{m+1} \int_{\Sigma_i\setminus K_i} e^2_{i,\delta} |\nabla^k (s - v^{\mathrm{Pal}})|^2 \mathrm{vol}_{\Sigma_i} + \|v\|^2,\end{aligned}\tag{4.8}$$

where $v^{\rm Pal}$ is defined in (3.8).

We next define a weighted Sobolev norm for the sections on Σ_T. Let s be a section of u^*TX with $s|_{\partial\Sigma_T} \in \Gamma(u^*TL)$ such that

$$s \in L^2_{m+1}((\Sigma_T, \partial\Sigma_T); u^*TX, u^*TL).$$

Since we take m large, s is continuous. So the value $s(0,1/2) \in T_{u(0,1/2)}X$ at $(0,1/2)$ is well defined.

We define the norm of s by

$$\begin{aligned}
\|s\|^2_{W^2_{m+1,\delta}} &= \sum_{k=0}^{m+1} \int_{K_1} |\nabla^k s|^2 \mathrm{vol}_{\Sigma_1} + \sum_{k=0}^{m+1} \int_{K_2} |\nabla^k s|^2 \mathrm{vol}_{\Sigma_2} \\
&\quad + \sum_{k=0}^{m+1} \int_{[-5T,5T]\times[0,1]} e^2_{T,\delta} |\nabla^k (s - s(0,1/2)^{\rm Pal})|^2 \mathrm{vol}_{\Sigma_i} \\
&\quad + \|s(0,1/2)\|^2.
\end{aligned} \tag{4.9}$$

For $\eta \in L^2_m((\Sigma_T, \partial\Sigma_T); u^*TX \otimes \Lambda^{0,1})$, we define

$$\|\eta\|^2_{L^2_{m,\delta}} = \sum_{k=0}^{m} \int_{\Sigma_T} e^2_{T,\delta} |\nabla^k \eta|^2 \mathrm{vol}_{\Sigma_1}. \tag{4.10}$$

These norms were used in [**FOOO1**, Section 7.1.3].

Remark 4.1. We remark that the norm $\|s\|_{W^2_{m+1,\delta}}$ is equivalent to the standard L^2_{m+1}-norm on $L^2_{m+1}((\Sigma_T, \partial\Sigma_T); u^*TX, u^*TL)$, when T is fixed. However to study its relationship with the norm (4.8) as $T \to \infty$, making this choice of T-dependent norm in this way is necessary. We remark that the ratio between the norm $\|s\|_{W^2_{m+1,\delta}}$ and the standard L^2_{m+1}-norm diverges as T goes to infinity.

For a subset W of Σ_i or Σ_T we define $\|s\|^2_{W^2_{m,\delta}(W\subset\Sigma_i)}$, $\|\eta\|^2_{L^2_{m,\delta}(W\subset\Sigma_T)}$ by restricting the domain of the integration (4.9) or (4.10) to W. We use this notation in order to specify the weight etc. we use. In case there is no possibility of confusion of the domain and weight etc. we use, we omit the domain Σ_i etc. from the notation $\|s\|^2_{W^2_{m,\delta}(\Sigma_i)}$ etc..

We also define the inner product on $W^2_{m+1,\delta}((\Sigma_i, \partial\Sigma_i); u_i^*TX, u_i^*TL)$ by setting

$$\begin{aligned}
\langle\!\langle (s_1, v_1), (s_2, v_2) \rangle\!\rangle_{L^2} &= \int_{\Sigma_i \setminus K_i} (s_1 - v_1^{\rm Pal}, s_2 - v_2^{\rm Pal}) \\
&\quad + \int_{K_i} (s_1, s_2) + (v_1, v_2)
\end{aligned} \tag{4.11}$$

for $(s_j, v_j) \in W^2_{m+1,\delta}((\Sigma_i, \partial\Sigma_i); u_i^*TX, u_i^*TL)$ for $j = 1, 2$.

CHAPTER 5

Proof of the gluing theorem II: Gluing by alternating method

We motivate the gluing construction we employ in this section by comparing it with the well-known Newton's iteration scheme of solving the equation $f(x) = 0$ for a real-valued function, starting from an approximate solution x_0 that is sufficiently close to a genuine solution nearby. The iteration scheme is the inductive scheme via the recurrence relation

$$x_{n+1} = x_n - \frac{f(x_n)}{f'(x_n)}. \tag{5.1}$$

Therefore

$$|x_{n+1} - x_n| = \left|\frac{f(x_n)}{f'(x_n)}\right| \tag{5.2}$$

and the expected solution is given by the infinite sum

$$x_\infty = x_0 + \sum_{n=1}^{\infty}(x_{n+1} - x_n)$$

provided the series converges. A simple way of ensuring the required convergence is to establish existence of constant $C > 0$, $0 \leq \mu < 1$ independent of n such that

$$\left|\frac{f(x_n)}{f'(x_n)}\right| \leq C\mu^n \tag{5.3}$$

for all n.

With this iteration scheme in our mind, we proceed with performing the scheme.

5.1. The first trial approximate solution

In our construction we obtain an approximate solution of the linearized equation (whose error corresponds to $f(x_0)/f'(x_0)$ in (5.1)) by an alternating method. We start with a pair of maps

$$u^\rho = (u_1^{\rho_1}, u_2^{\rho_2}) \in \mathcal{M}^{\mathcal{E}_1}((\Sigma_1, \vec{z}_1); u_1)_{\epsilon_2} \times_L \mathcal{M}^{\mathcal{E}_2}((\Sigma_2, \vec{z}_2); u_2)_{\epsilon_2}$$

for each sufficiently small $\epsilon_2 > 0$. Here $\rho_i \in V_i$ and $u_i^{\rho_i}$ is the corresponding map $(\Sigma_i, \partial\Sigma_i) \to (X, L)$ given by (3.26). Let $\rho = (\rho_1, \rho_2)$. We put

$$p_0^\rho = \lim_{\tau\to\infty} u_1^{\rho_1}(\tau, t) = \lim_{\tau\to-\infty} u_2^{\rho_2}(\tau, t).$$

Pregluing: This step corresponds to picking the initial trial approximate solution x_0 in Newton's scheme above.

Step 0-i (Approximate solutions):

DEFINITION 5.1. We define

$$u^{\rho}_{T,(0)} = \begin{cases} \operatorname{Exp}\left(p^{\rho}_0, \chi^{\leftarrow}_{\mathcal{B}} \mathrm{E}(p^{\rho}_0, u^{\rho_1}_1) + \chi^{\rightarrow}_{\mathcal{A}} \mathrm{E}(p^{\rho}_0, u^{\rho_2}_2)\right) & \text{on } [-5T, 5T] \times [0,1] \\ u^{\rho_1}_1 & \text{on } K_1 \\ u^{\rho_2}_2 & \text{on } K_2. \end{cases} \tag{5.4}$$

This pre-glued map $u^{\rho}_{T,(0)}$ plays the role of x_0 in the Newton scheme for solving the equation

$$\Pi^{\perp}_{\mathcal{E}_T} \overline{\partial} u = 0 \tag{5.5}$$

on $L^2_{m+1,\delta}(\Sigma_T, \partial\Sigma_T; (u^{\rho}_{T,(0)})^*TX, (u^{\rho}_{T,(0)})^*TL)$ over Σ_T, where $\Pi^{\perp}_{\mathcal{E}_T}$ is the projection to a subspace $\mathcal{E}^{\perp}_T$. We will need to take the derivative of the map $u \mapsto \Pi^{\perp}_{\mathcal{E}_T} \overline{\partial} u$ the analog to $f'(x_n)$ in the Newton's scheme. We will actually use an approximate derivative thereof whose explanation is now in order.

Step 0-ii (Approximate linearization): Note that the subspace $\mathcal{E}_i(u'_i)$ varies on u'_i. (See (3.18) for the definition.) So to obtain a linearized equation of (5.5) at the given map u'_i, we need to take this dependence into account.

Let $\Pi_{\mathcal{E}_i(u'_i)}$ denote the L^2-orthogonal projection of $L^2_{m,\delta}(\Sigma_i; (u'_i)^*TX \otimes \Lambda^{0,1})$ to $\mathcal{E}_i(u'_i)$ or its associated idempotent as an endomorphism

$$\Pi_{\mathcal{E}_i(u'_i)} : L^2_{m,\delta}(\Sigma_i; (u'_i)^*TX \otimes \Lambda^{0,1}) \to L^2_{m,\delta}(\Sigma_i; (u'_i)^*TX \otimes \Lambda^{0,1}).$$

More explicitly, we represent the projection as follows: we choose a basis $\{\mathbf{e}_{i,a}\}$ of $\mathcal{E}_i$ and set

$$\mathbf{e}'_{i,a}(u'_i) := I_{u'_i}(\mathbf{e}_{i,a}), \quad a = 1, \dots, \dim \mathcal{E}_i(u'_i).$$

We apply the standard Gram-Schmidt process to $\mathbf{e}'_{i,a}(u'_i)$ and denote by $\mathbf{e}_{i,a}(u'_i)$ the resulting orthonormal basis of $\mathcal{E}_i(u'_i)$. Then, for $A \in L^2_{m,\delta}(\Sigma_i; (u'_i)^*TX \otimes \Lambda^{0,1})$ we put

$$\Pi_{\mathcal{E}_i(u'_i)}(A) = \sum_{a=1}^{\dim E_i} \langle\!\langle A, \mathbf{e}_{i,a}(u'_i)\rangle\!\rangle_{L^2(K_1)} \mathbf{e}_{i,a}(u'_i), \tag{5.6}$$

where $\mathbf{e}_{i,a}(u'_i)$ are supported in K_i.

For given $V \in \Gamma((\Sigma_i, \partial\Sigma_i), (u'_i)^*TX, (u'_i)^*TL)$, we put

$$(D_{u'_i}\mathcal{E}_i)(A, V) := \left.\frac{d}{ds}\right|_{s=0} (\mathrm{Pal}^{(0,1)}_i)^{-1} \Pi_{\mathcal{E}_i(\operatorname{Exp}(u'_i, sV))}(\mathrm{Pal}^{(0,1)}_i(A)). \tag{5.7}$$

Here $\mathrm{Pal}^{(0,1)}_i$ is the closure of the map $(\mathrm{Pal}^{\operatorname{Exp}(u'_i, sV)}_{u'_i})^{(0,1)}$ defined in (2.7). The right hand side is well defined in view of (5.6). We then put

$$\Pi^{\perp}_{\mathcal{E}_i(u'_i)}(A) = A - \Pi_{\mathcal{E}_i(u'_i)}(A).$$

The equation (3.22) is equivalent to $\Pi^{\perp}_{\mathcal{E}_i(u'_i)}(\overline{\partial}) = 0$. By evaluating the derivative

$$\left.\frac{\partial}{\partial s}\right|_{s=0} (\mathrm{Pal}^{(0,1)}_i)^{-1} \left(\Pi^{\perp}_{\mathcal{E}_i(\operatorname{Exp}(u'_i, sV))} \overline{\partial} \operatorname{Exp}(u'_i, sV)\right)$$

we obtain the linearization operator of (3.22).

As an operator into $L^2_{m,\delta}(\Sigma_i; (u'_i)^*TX \otimes \Lambda^{0,1})$ it becomes

$$V \mapsto (\Pi^{\perp}_{\mathcal{E}_i(u'_i)} \circ D_{u'_i}\overline{\partial})(V) - (D_{u'_i}\mathcal{E}_i)(\overline{\partial} u'_i, V).$$

We define:

$$(D_{u'_i}\overline{\partial})^{\perp}(V) = (D_{u'_i}\overline{\partial})(V) - (D_{u'_i}\mathcal{E}_i)(\overline{\partial}u'_i, V). \tag{5.8}$$

We use a variant of this operator as an approximate derivative of the map $u \mapsto \Pi^{\perp}_{\mathcal{E}_i(u)}\overline{\partial}u$. See (5.20).

Step 0-3 (Error estimates): We recall that $\text{supp}(\mathcal{E}_i^{\mathfrak{ob}}) \subset \text{Int}\, K_i$ by the definition of $\mathcal{E}_i^{\mathfrak{ob}}$.

LEMMA 5.2. *There exists $\mathfrak{e}^{\rho}_{i,T,(0)} \in \mathcal{E}_i^{\mathfrak{ob}}$ such that*

$$\|\overline{\partial}u^{\rho}_{T,(0)} - \mathfrak{e}^{\rho}_{1,T,(0)} - \mathfrak{e}^{\rho}_{2,T,(0)}\|_{L^2_{m,\delta}} < C_{1,m}e^{-\delta_1 T}. \tag{5.9}$$

Here $C_{1,m}$ is a constant depending on X, L, u_i and m and $u^{\rho}_{T,(0)}$ is as in Definition 5.1.

Moreover for any $\epsilon(4) > 0$ there exists $\epsilon_{\epsilon(4),(5.10)}$ such that

$$\|\mathfrak{e}^{\rho}_{i,T,(0)}\|_{L^2(K_i)} < \epsilon(4) \tag{5.10}$$

if $\rho = (\rho_1, \rho_2) \in V_1(\epsilon) \times_L V_2(\epsilon)$ with $\epsilon < \epsilon_{\epsilon(4).(5.10)}$.

Furthermore $\mathfrak{e}^{\rho}_{i,T,(0)}$ is supported in K_i and is independent of T as an element of $L^2_m(K_i)$.

In (5.9) we regard $\mathfrak{e}^{\rho}_{i,T,(0)} \in \mathcal{E}_i^{\mathfrak{ob}}$ as an element of $\mathcal{E}_i(u^{\rho}_{T,(0)})$ by (3.18) with $u' = u^{\rho}_{T,(0)}$.

REMARK 5.3. Note throughout the discussion of Chapters 3-7 we fix X, L, u_i. So we do not mention dependence of constants on them hereafter.

PROOF. By definition, $\overline{\partial}u_i^{\rho} \in \mathcal{E}_i(u_i^{\rho})$. We put

$$\mathfrak{e}^{\rho}_{i,T,(0)} := \overline{\partial}u_i^{\rho} \in \mathcal{E}_i(u_i^{\rho}) \cong \mathcal{E}_i^{\mathfrak{ob}} \tag{5.11}$$

extended to be zero to $\Sigma_T \backslash K_i$. Then, by definition, the support of $\overline{\partial}u^{\rho}_{T,(0)} - \mathfrak{e}^{\rho}_{1,T,(0)} - \mathfrak{e}^{\rho}_{2,T,(0)}$ is contained in $[-2T, 2T] \times [0,1]$. In fact on $K_1 \cup [-5T, -2T] \times [0,1]$, $u^{\rho}_{T,(0)} = u_1^{\rho_1}$ and on $K_2 \cup [2T, 5T] \times [0,1]$, $u^{\rho}_{T,(0)} = u_2^{\rho_2}$, so $\overline{\partial}u^{\rho}_{T,(0)} = \mathfrak{e}^{\rho}_{1,T,(0)} + \mathfrak{e}^{\rho}_{2,T,(0)}$, thereon. Moreover by (2.10) and (5.4) the C^k norm of the map $(\tau, t) \mapsto \text{E}(p_0^{\rho}, u^{\rho}_{T,(0)}(\tau, t)) : [-2T, 2T] \times [0,1] \to T_{p_0^{\rho}}X$ at (τ, t) is smaller than $C_k e^{-\delta_1 d(\partial[-5T,5T],\tau)} \le C_k e^{-3\delta_1 T}$, for any k. Therefore we obtain the estimate (5.9). □

REMARK 5.4. Note (5.11) (resp. (5.58), (C.12)) specifies the choice of $\mathfrak{e}^{\rho}_{i,T,(0)}$ (resp. $\mathfrak{e}^{\rho}_{i,T,(1)}$, $\mathfrak{e}^{\rho}_{i,T,(\kappa)}$). This is the choice taken in this proof.

Step 0-4 (Separating error terms into two parts): We denote the initial error by

$$\text{Err}^{\rho}_{T,(0)} = \overline{\partial}u^{\rho}_{T,(0)} - \mathfrak{e}^{\rho}_{1,T,(0)} - \mathfrak{e}^{\rho}_{2,T,(0)}. \tag{5.12}$$

DEFINITION 5.5. We put

$$\begin{aligned}
\text{Err}^{\rho}_{1,T,(0)} &= \chi_{\mathcal{X}}^{\leftarrow}(\text{Err}^{\rho}_{T,(0)}) = \chi_{\mathcal{X}}^{\leftarrow}(\overline{\partial}u^{\rho}_{T,(0)} - \mathfrak{e}^{\rho}_{1,T,(0)}),\\
\text{Err}^{\rho}_{2,T,(0)} &= \chi_{\mathcal{X}}^{\rightarrow}(\text{Err}^{\rho}_{T,(0)}) = \chi_{\mathcal{X}}^{\rightarrow}(\overline{\partial}u^{\rho}_{T,(0)} - \mathfrak{e}^{\rho}_{2,T,(0)}).
\end{aligned}$$

We regard them as elements of the weighted Sobolev spaces $L^2_{m,\delta}(\Sigma_1;(u_1^\rho)^*TX\otimes\Lambda^{0,1})$ and $L^2_{m,\delta}(\Sigma_2;(u_2^\rho)^*TX\otimes\Lambda^{0,1})$ respectively. (We extend them by 0 outside $\operatorname{supp}\chi_{\mathcal X}^{\leftarrow}$, $\operatorname{supp}\chi_{\mathcal X}^{\rightarrow}$, respectively.)

5.2. The first iteration

We now start the first step of iteration.

Step 1-1 (Approximate solution for linearization): This step corresponds to solving the linearized equation

$$f'(x_0)(v_0)=f(x_0),\qquad \text{namely } v_0=\frac{f(x_0)}{f'(x_0)} \tag{5.13}$$

in Newton's iteration scheme.

We first cut-off $u^\rho_{T,(0)}$ and extend it to obtain maps $\hat u^\rho_{i,T,(0)}:(\Sigma_i,\partial\Sigma_i)\to(X,L)$ ($i=1,2$) as follows. (This map is used to set the linearized operator (5.15).)

(5.14)

$$\hat u^\rho_{1,T,(0)}(z)=\begin{cases}\operatorname{Exp}\left(p_0^\rho,\chi_{\mathcal B}^{\leftarrow}(\tau-T,t)\operatorname{E}(p_0^\rho,u^\rho_{T,(0)}(\tau,t))\right) & \text{if } z=(\tau,t)\in[-5T,5T]\times[0,1]\\ u^\rho_{T,(0)}(z) & \text{if } z\in K_1\\ p_0^\rho & \text{if } z\in[5T,\infty)\times[0,1].\end{cases}$$

$$\hat u^\rho_{2,T,(0)}(z)=\begin{cases}\operatorname{Exp}\left(p_0^\rho,\chi_{\mathcal A}^{\rightarrow}(\tau+T,t)\operatorname{E}(p_0^\rho,u^\rho_{T,(0)}(\tau,t))\right) & \text{if } z=(\tau,t)\in[-5T,5T]\times[0,1]\\ u^\rho_{T,(0)}(z) & \text{if } z\in K_2\\ p_0^\rho & \text{if } z\in(-\infty,-5T]\times[0,1].\end{cases}$$

Now we let

$$\begin{aligned}D_{\hat u^\rho_{i,T,(0)}}\overline\partial : W^2_{m+1,\delta}&\left((\Sigma_i,\partial\Sigma_i);(\hat u^\rho_{i,T,(0)})^*TX,(\hat u^\rho_{i,T,(0)})^*TL\right)\\ &\to L^2_{m,\delta}\left(\Sigma_i;(\hat u^\rho_{i,T,(0)})^*TX\otimes\Lambda^{0,1}\right)\end{aligned}\tag{5.15}$$

be the (covariant) linearization of the Cauchy-Riemann equation at $\hat u^\rho_{i,T,(0)}$, i.e.,

$$(D_{\hat u^\rho_{i,T,(0)}}\overline\partial)(V)=\left.\frac{\partial}{\partial s}\right|_{s=0}\left((\mathrm{Pal}_i^{(0,1)})^{-1}(\overline\partial\operatorname{Exp}(\hat u^\rho_{i,T,(0)},sV))\right).\tag{5.16}$$

Here $\mathrm{Pal}_i^{(0,1)}$ is the $(0,1)$ part of the parallel translation along the map $r\mapsto\operatorname{Exp}(\hat u^\rho_{i,T,(0)},rV)$, $r\in[0,s]$ for some $s\in[0,1]$.

LEMMA 5.6. *There exist $\epsilon_2, T_{(5.18)}$ with the following properties. Let $\mathcal E_i^{\mathfrak{ob}}$ be the subspace chosen in* (3.14). *Define*

$$\mathcal E_i(\hat u^\rho_{i,T,(0)}):=I_{\hat u^\rho_{i,T,(0)}}(\mathcal E_i^{\mathfrak{ob}})\tag{5.17}$$

as the subspace of $L^2_{m,\delta}\left(\Sigma_i;(\hat u^\rho_{i,T,(0)})^*TX\otimes\Lambda^{0,1}\right)$. Then the following holds if $T>T_{(5.18)}$ and $\rho\in V_1(\epsilon)\times_L V_2(\epsilon)$ with $\epsilon<\epsilon_2$:

$$\operatorname{Im}(D_{\hat u^\rho_{i,T,(0)}}\overline\partial)+\mathcal E_i(\hat u^\rho_{i,T,(0)})=L^2_{m,\delta}\left(\Sigma_i;(\hat u^\rho_{i,T,(0)})^*TX\otimes\Lambda^{0,1}\right).\tag{5.18}$$

Moreover

$$(5.19)\qquad \begin{aligned}&\mathrm{Dev}_{1,\infty} - \mathrm{Dev}_{2,\infty} v\\ &: (D_{\hat u^\rho_{1,T,(0)}}\overline{\partial})^{-1}(\mathcal{E}_1(\hat u^\rho_{1,T,(0)})) \oplus (D_{\hat u^\rho_{2,T,(0)}}\overline{\partial})^{-1}(\mathcal{E}_2(\hat u^\rho_{2,T,(0)})) \to T_{p^\rho} L\end{aligned}$$

is surjective.

PROOF. Since $\hat u^\rho_{i,T,(0)}$ is close to u_i in an exponential order, this is a consequence of Assumption 3.12. □

We note that $\overline{\partial}\hat u^\rho_{i,T,(0)} - \mathfrak{e}^\rho_{i,T,(0)}$ is exponentially small. Therefore the map

$$V \mapsto (D_{\hat u^\rho_{i,T,(0)}}\overline{\partial})(V) - (D_{\hat u^\rho_{i,T,(0)}}\mathcal{E}_i)(\mathfrak{e}^\rho_{i,T,(0)}, V),$$

which is obtained by replacing $\overline{\partial}\hat u^\rho_{i,T,(0)}$ by $\mathfrak{e}^\rho_{i,T,(0)}$, is exponentially close to the linearization operator $(D_{\hat u^\rho_{i,T,(0)}}\overline{\partial})^\perp$, defined in (5.8). We will consider this approximate linearization instead and denote it by

$$D^{\mathrm{app},(0)}_{\hat u^\rho_{i,T,(0)}} : W^2_{m+1,\delta}(\Sigma_i; (\hat u^\rho_{i,T,(0)})^*TX) \to L^2_{m,\delta}(\Sigma_i; (\hat u^\rho_{i,T,(0)})^*TX \otimes \Lambda^{0,1})$$

with

$$(5.20)\qquad D^{\mathrm{app},(0)}_{\hat u^\rho_{i,T,(0)}}(V) := (D_{\hat u^\rho_{i,T,(0)}}\overline{\partial})(V) - (D_{\hat u^\rho_{i,T,(0)}}\mathcal{E}_i)(\mathfrak{e}^\rho_{i,T,(0)}, V).$$

LEMMA 5.7. *There exist $T_{(5.21)} > 0$ and $\epsilon_2 > 0$ with the following properties. If $T > T_{(5.21)}$ and $\rho \in V_1(\epsilon) \times_L V_2(\epsilon)$ with $\epsilon < \epsilon_2$ then*

$$(5.21)\qquad \mathit{Im}\, D^{app,(0)}_{\hat u^\rho_{i,T,(0)}} + \mathcal{E}_i(\hat u^\rho_{i,T,(0)}) = L^2_{m,\delta}(\Sigma_i; (\hat u^\rho_{i,T,(0)})^*TX \otimes \Lambda^{0,1}).$$

Moreover

$$(5.22)\qquad \mathrm{Dev}_{1,\infty} - \mathrm{Dev}_{2,\infty} : (D^{app,(0)}_{\hat u^\rho_{1,T,(0)}})^{-1}(\mathcal{E}_1) \oplus (D^{app,(0)}_{\hat u^\rho_{2,T,(0)}})^{-1}(\mathcal{E}_2) \to T_{p^\rho} L$$

is surjective. Here we put $\mathcal{E}_i = \mathcal{E}_i(\hat u^\rho_{i,T,(0)})$.

PROOF. The inequality (5.10) implies that $(D_{\hat u^\rho_{1,T,(0)}}\mathcal{E}_1)(\mathfrak{e}^\rho_{1,T,(0)}, \cdot)$ is small in the operator norm. The lemma then follows from Lemma 5.6. □

REMARK 5.8. Recall that

$$V_i(\epsilon) \subset \mathcal{M}^{\mathcal{E}_i}((\Sigma_i, \vec z_i); \beta_i)_{\epsilon_2}$$

consists of smooth maps. Note that (5.10) and conclusions of Lemmata 5.6 and 5.7 are proved by taking a small neighborhood $V_i(\epsilon)$ of 0 (in $\tilde V_i$) with respect to the C^m norm.

However we can take $V_i(\epsilon)$ so that it becomes independent of m and the conclusion of Lemma 5.7 holds for all m. In fact the elliptic regularity implies that if the conclusion of Lemma 5.7 holds for some m then it holds for all $m' > m$. The inequality (5.10) holds for that particular m only. Only this inequality for m is used to show Lemma 5.7. In other words, ϵ_2 is independent of m.

Based on this remark, we fix $V_1 = V_1(\epsilon_2), V_2 = V_2(\epsilon_2)$ now and will never change them in Chapters 5, 6. In other words, the constant ϵ_2 is fixed at this stage.

We consider the finite dimensional subspace

$$(5.23)\qquad \mathrm{Ker}(\mathrm{Dev}_{1,\infty} - \mathrm{Dev}_{2,\infty}) \cap (((D_{u_1}\overline{\partial})^\perp)^{-1}(\mathcal{E}_1) \oplus ((D_{u_2}\overline{\partial})^\perp)^{-1}(\mathcal{E}_2))$$

of

$$\operatorname{Ker}(Dev_{1,\infty} - Dev_{2,\infty}) \cap \bigoplus_{i=1}^{2} W^2_{m+1,\delta}((\Sigma_i,\partial\Sigma_i); u_i^*TX, (u_i^*TL)) \tag{5.24}$$

which consists of smooth sections. Here we recall $\mathcal{E}_i = \mathcal{E}_i(u_i)$. Note $\overline{\partial} u_i = 0$. Therefore $(D_{u_i}\overline{\partial})^{\perp} = D_{u_i}\overline{\partial}$ by (5.8).

DEFINITION 5.9. We define a linear subspace $\mathfrak{H}(\mathcal{E}_1,\mathcal{E}_2)$ of (5.24) by

$$\mathfrak{H}(\mathcal{E}_1,\mathcal{E}_2) \perp (5.23), \qquad \mathfrak{H}(\mathcal{E}_1,\mathcal{E}_2) \oplus (5.23) = (5.24). \tag{5.25}$$

In other words, it is the L^2 orthogonal complement of (5.23) in (5.24). Here the L^2 inner product is defined by (4.11).

Note the supports of elements of $\mathcal{E}_i(u_i)$ are contained in K_i. Therefore the projection of C^k (resp. L^2_k) section to $\mathfrak{H}(\mathcal{E}_1,\mathcal{E}_2)$ is C^k (resp. L^2_k).

We use the parallel transport to define an isomorphism from

$$W^2_{m+1,\delta}((\Sigma_i,\partial\Sigma_i); u_i^*TX, u_i^*TL) \tag{5.26}$$

to

$$W^2_{m+1,\delta}((\Sigma_i,\partial\Sigma_i); (\hat u^{\rho}_{i,T,(0)})^*TX, (\hat u^{\rho}_{i,T,(0)})^*TL), \tag{5.27}$$

as follows.

DEFINITION 5.10. We define the isomorphism $\Phi_{i;(0)}(\rho,T) : (5.26) \to (5.27)$ to be the closure of the map $\mathrm{Pal}_{u_i}^{\hat u^{\rho}_{i,T,(0)}}$ given in (2.6). With a slight abuse of notation, we still denote

$$\Phi_{i;(0)}(\rho,T) = \mathrm{Pal}_{u_i}^{\hat u^{\rho}_{i,T,(0)}}. \tag{5.28}$$

Here the right hand side is the closure of the map $\mathrm{Pal}_{u_i}^{\hat u^{\rho}_{i,T,(0)}}$ in (2.6).

LEMMA 5.11.

(1) *$\Phi_{i;(0)}(\rho,T)$ is an isomorphism between* (5.26) *and* (5.27). .
(2) *The linear map $(\Phi_{1;(0)}(\rho,T),\Phi_{2;(0)}(\rho,T))$ sends the subspace* (5.23) *into the subspace*

$$\begin{aligned} &\operatorname{Ker}(Dev_{1,\infty} - Dev_{2,\infty}) \\ &\cap \bigoplus_{i=1}^{2} W^2_{m+1,\delta}((\Sigma_i,\partial\Sigma_i); (\hat u^{\rho}_{i,T,(0)})^*TX, (\hat u^{\rho}_{i,T,(0)})^*TL). \end{aligned} \tag{5.29}$$

With Lemma 5.7 the proof is easy and is omitted.

DEFINITION 5.12. We denote by $\mathfrak{H}_{(0)}(\mathcal{E}_1,\mathcal{E}_2;\rho,T)$ the image of the restriction of the linear map $(\Phi_{1,(0)}(\rho,T),\Phi_{2,(0)}(\rho,T))$ to $\mathfrak{H}(\mathcal{E}_1,\mathcal{E}_2)$. It is a subspace of (5.29).

DEFINITION 5.13. Let $\mathrm{Err}^{\rho}_{i,T,(0)}$ be the functions defined in Definition 5.5. We define the triple $(V^{\rho}_{T,1,(1)}, V^{\rho}_{T,2,(1)}, \Delta p^{\rho}_{T,(1)})$ to be the unique solution satisfying the requirements

$$D^{\mathrm{app},(0)}_{\hat u^{\rho}_{i,T,(0)}}(V^{\rho}_{T,i,(1)}) + \mathrm{Err}^{\rho}_{i,T,(0)} \in \mathcal{E}_i(\hat u^{\rho}_{i,T,(0)}), \tag{5.30}$$

and

$$((V^{\rho}_{T,1,(1)}, \Delta p^{\rho}_{T,(1)}), (V^{\rho}_{T,2,(1)}, \Delta p^{\rho}_{T,(1)})) \in \mathfrak{H}_{(0)}(\mathcal{E}_1,\mathcal{E}_2;\rho,T). \tag{5.31}$$

Lemma 5.11 implies that such $(V^{\rho}_{T,1,(1)}, V^{\rho}_{T,2,(1)}, \Delta p^{\rho}_{T,(1)})$ exists and is unique if $T > T_{(5.33)}$ with the sufficiently large $T_{(5.33)}$ and ρ is in $V_1 \times_L V_2$ chosen at Remark 5.8. In fact, when we consider the subspace

$$\mathrm{Ker}(Dev_{1,\infty} - Dev_{2,\infty}) \cap \bigoplus_{i=1}^{2} (D^{\mathrm{app},(0)}_{\hat{u}^{\rho}_{i,T,(0)}})^{-1}(\mathcal{E}_i) \tag{5.32}$$

of (5.29) we still have a direct sum decomposition

$$(5.32) \oplus \mathfrak{H}_{(0)}(\mathcal{E}_1, \mathcal{E}_2; \rho, T) = (5.29). \tag{5.33}$$

Since (5.33) holds for $T = \infty$ by definition, so it holds for $T > T_{(5.33)}$.

The unique solution $(V^{\rho}_{T,1,(1)}, V^{\rho}_{T,2,(1)}, \Delta p^{\rho}_{T,(1)})$ corresponds to the solution v_0 to (5.13) in Newton's scheme. Then the following estimate corresponds to estimating $|x_1 - x_0| = |v_0|$ for the solution v_0 to the linearized equation (5.13) in Newton's scheme.

REMARK 5.14. Instead of $\mathfrak{H}_{(0)}(\mathcal{E}_1, \mathcal{E}_2; \rho, T)$ we can use an L^2 orthogonal complement of (5.32) in (5.29). This choice looks more natural and actually works for the proof of Theorems 3.13 and 6.4. We take the current choice since when we study T and ρ derivative in Section 6.3, it makes calculation slightly shorter.

LEMMA 5.15. *There exist $C_{2,m} > 0, T_{m,(5.34)} > 0$ such that if $T > T_{m,(5.34)}$, then*

$$\|(V^{\rho}_{T,i,(1)}, \Delta p^{\rho}_{T,(1)})\|_{W^2_{m+1,\delta}(\Sigma_i)} \le C_{2,m} e^{-\delta_1 T}. \tag{5.34}$$

This is immediate from construction and uniform boundedness of the right inverse of the operator

$$\Pi^{\perp}_{\mathcal{E}_i} \circ D^{\mathrm{app},(0)}_{\hat{u}^{\rho}_{i,T,(0)}} = \Pi^{\perp}_{\mathcal{E}_i} \circ (D_{\hat{u}^{\rho}_{i,T,(0)}} \overline{\partial} - (D_{\hat{u}^{\rho}_{i,T,(0)}} \mathcal{E}_i)(\mathfrak{e}^{\rho}_{i,T,(0)}, \cdot))$$

and the exponential decay estimates of the error term. (Lemma 5.2.)

Step 1-2 (Gluing solutions): We use $(V^{\rho}_{T,1,(1)}, V^{\rho}_{T,2,(1)}, \Delta p^{\rho}_{T,(1)})$ to find an approximate solution $u^{\rho}_{T,(1)}$ of the next level. This corresponds to writing down

$$x_1 = x_0 - v_0 = x_0 - \frac{f(x_0)}{f'(x_0)}$$

in Newton's scheme.

DEFINITION 5.16. We define $u^{\rho}_{T,(1)}(z)$ as follows.

(1) If $z \in K_1$, we put

$$u^{\rho}_{T,(1)}(z) = \mathrm{Exp}(\hat{u}^{\rho}_{1,T,(0)}(z), V^{\rho}_{T,1,(1)}(z)). \tag{5.35}$$

(2) If $z \in K_2$, we put

$$u^{\rho}_{T,(1)}(z) = \mathrm{Exp}(\hat{u}^{\rho}_{2,T,(0)}(z), V^{\rho}_{T,2,(1)}(z)). \tag{5.36}$$

(3) If $z = (\tau, t) \in [-5T, 5T] \times [0,1]$, we put

$$\begin{aligned} u^{\rho}_{T,(1)}(\tau,t) = \mathrm{Exp}\Big(& u^{\rho}_{T,(0)}(\tau,t), \chi^{\leftarrow}_{\mathcal{B}}(\tau,t)(V^{\rho}_{T,1,(1)}(\tau,t) - (\Delta p^{\rho}_{T,(1)})^{\mathrm{Pal}}) \\ &+ \chi^{\rightarrow}_{\mathcal{A}}(\tau,t)(V^{\rho}_{T,2,(1)}(\tau,t) - (\Delta p^{\rho}_{T,(1)})^{\mathrm{Pal}}) + (\Delta p^{\rho}_{T,(1)})^{\mathrm{Pal}}\Big). \end{aligned} \tag{5.37}$$

We recall that $\hat u^{\rho}_{1,T,(0)}(z) = u^{\rho}_{T,(0)}(z)$ on K_1 and $\hat u^{\rho}_{2,T,(0)}(z) = u^{\rho}_{T,(0)}(z)$ on K_2.

Step 1-3 (Error estimates): Let a constant $0 < \mu < 1$ be given, which we fix throughout the rest of the proof. For example we can take $\mu = 1/2$.

This step corresponds to establishing the inequality (5.3) for $n = 1$ in Newton's iteration scheme.

PROPOSITION 5.17. *There exists $C_{3,m}$ such that for any $\epsilon(4) > 0$, we can choose $T_{m,\epsilon(4),(5.38)} > 0$ with the following properties: there exist $\mathfrak{e}^{\rho}_{i,T,(1)} \in \mathcal{E}_i$ that satisfies*

$$\|\mathrm{Err}^{\rho}_{T,(1)}\|_{L^2_{m,\delta}} < C_{1,m}\epsilon(4)\mu e^{-\delta_1 T} \tag{5.38}$$

and

$$\|\mathfrak{e}^{\rho}_{i,T,(1)}\|_{L^2_m(K_i)} < C_{3,m}e^{-\delta_1 T}. \tag{5.39}$$

for all $T > T_{m,(5.38)}$. (Here $C_{1,m}$ is the constant given in Lemma 5.2.)

In (5.38) *we denote*

$$\mathrm{Err}^{\rho}_{T,(1)} = \overline{\partial} u^{\rho}_{T,(1)} - \sum_{a=0}^{1} \mathfrak{e}^{\rho}_{1,T,(a)} - \sum_{a=0}^{1} \mathfrak{e}^{\rho}_{2,T,(a)}. \tag{5.40}$$

Note Remark 5.4 applies here.

PROOF. The existence of $\mathfrak{e}^{\rho}_{i,T,(1)}$ satisfying

$$\begin{aligned}&\|\overline{\partial} u^{\rho}_{T,(1)} - (\mathfrak{e}^{\rho}_{1,T,(0)} + \mathfrak{e}^{\rho}_{1,T,(1)}) - (\mathfrak{e}^{\rho}_{2,T,(0)} + \mathfrak{e}^{\rho}_{2,T,(1)})\|_{L^2_{m,\delta}(K_1\cup K_2 \subset \Sigma_T)}\\ &< C_{1,m}\epsilon(4)\mu e^{-\delta_1 T}/10\end{aligned}$$

is a consequence of the fact that (5.8) is the linearized equation of (3.19) and the estimate (5.34). More explicitly, we can prove this by a standard quadratic estimate whose details are now in order.

We do the necessary estimates on the regions K_1, K_2, $([-5T, 5T]\setminus[-T-1, T+1]) \times [0,1]$ and region $[-T-1, T+1] \times [0,1]$ separately.

Step 1-3-1: Estimate on K_i and Definition of $\mathfrak{e}^{\rho}_{i,T,(1)}$

We first estimate on K_1. We put $K_1^+ = K_1 \cup [-5T, -5T+1] \times [0,1]$.

REMARK 5.18. The estimate on K_1 is rather lengthy. However it consists only of standard and straightforward calculation. This estimate is *not* a part of the main idea of the proof. The main idea of the proof of this paper lies in the points mentioned in Remarks 5.21 and 6.13.

During this estimate on K_1 and on $[-5T, -T-1] \times [0,1]$ we use the following simplified notation.

$$\begin{aligned} u &= \hat u^{\rho}_{1,T,(0)}, & V &= V^{\rho}_{T,1,(1)},\\ \mathcal{P} &= (\mathrm{Pal}_1^{(0,1)})^{-1}, & \mathfrak{e} &= \mathfrak{e}^{\rho}_{1,T,(0)}.\end{aligned} \tag{5.41}$$

Here $\mathrm{Pal}_1^{(0,1)}$ is the $(0,1)$ part of the parallel translation along the curve $s \mapsto \mathrm{Exp}(u, sV)$, $s \in [0, s_0]$ for $s_0 \in [0,1]$.

By Fundamental Theorem of Calculus

$$g(1) = g(0) + \int_0^1 g'(s)\,ds = g(0) + g'(0) + \int_0^1 ds\left(\int_0^s g''(r)\,dr\right) \tag{5.42}$$

applied to the function g which is valued in $L^2_{m,\delta}(K_1^+; (\hat{u}^\rho_{1,T,(0)})^*TX \otimes \Lambda^{0,1})$ and is given by $g(s) = \mathcal{P}\overline{\partial}\,(\mathrm{Exp}(u, sV))$, we derive

$$
\begin{aligned}
&\mathcal{P}\overline{\partial}(\mathrm{Exp}(u, V)) \\
&= \overline{\partial}(\mathrm{Exp}(u, 0)) + \int_0^1 \frac{\partial}{\partial s}\left(\mathcal{P}\overline{\partial}(\mathrm{Exp}(u, sV))\right) ds \\
&= \overline{\partial} u + (D_u\overline{\partial})(V) + \int_0^1 ds \int_0^s \left(\frac{\partial}{\partial r}\right)^2 (\mathcal{P}\overline{\partial}(\mathrm{Exp}(u, rV))\, dr
\end{aligned}
\tag{5.43}
$$

on $L^2_{m,\delta}(\Sigma_T; u^*TX \otimes \Lambda^{0,1})$.

On the other hand, we have the following estimate.

LEMMA 5.19. *Let $m \geq 2$. Then*

$$
\begin{aligned}
\left\| \int_0^1 ds \int_0^s \left(\frac{\partial}{\partial r}\right)^2 (\mathcal{P}\overline{\partial}(\mathrm{Exp}(u, rV))\, dr \right\|_{L^2_m(K_1)} &\leq C_{m,(5.44)} \|V\|^2_{L^2_{m+1,\delta}(K_1^+)} \\
&\leq C'_{m,(5.44)} e^{-2\delta_1 T}.
\end{aligned}
\tag{5.44}
$$

PROOF. The second inequality follows from (5.34). On the other hand, the first one is rather an immediate consequence of the fact that the covariant derivative $\nabla_r^2 \mathcal{P} = \left(\frac{\partial}{\partial r}\right)^2 \mathcal{P}$ do not differentiate V. Especially this inequality is obvious for the case of flat metric and the standard complex structure on $\mathbb{C}^n$. For the readers' convenience, we provide a full explanation on why the presence of exponential maps and the parallel transports do not hinder obtaining the required estimate in Appendix A. □

We also have

$$
\begin{aligned}
&\mathcal{P} \circ \Pi^\perp_{\mathcal{E}_1(\mathrm{Exp}(u,V))} \circ \mathcal{P}^{-1} \\
&= \Pi^\perp_{\mathcal{E}_1(u)} + \int_0^1 \frac{d}{ds}\left(\mathcal{P} \circ \Pi^\perp_{\mathcal{E}_1(\mathrm{Exp}(u,sV)} \circ \mathcal{P}^{-1}\right) ds \\
&= \Pi^\perp_{\mathcal{E}_1(u)} - (D_u\mathcal{E}_1)(\cdot, V) + \int_0^1 ds \int_0^s \frac{d^2}{dr^2}\left(\mathcal{P} \circ \Pi^\perp_{\mathcal{E}_1(\mathrm{Exp}(u,rV))} \circ \mathcal{P}^{-1}\right) dr.
\end{aligned}
\tag{5.45}
$$

(See (5.7) for the definition of $D_u\mathcal{E}_1$.)

We can estimate the third term of the right hand side of (5.45) in the same way as in (5.44) and also get

$$
\left\| \int_0^1 ds \int_0^s \frac{d^2}{dr^2}\left(\mathcal{P} \circ \Pi^\perp_{\mathcal{E}_1(\mathrm{Exp}(u,rV))} \circ \mathcal{P}^{-1}\right) dr \right\|_{L^2_m(K_1)} \leq C_{m,(5.46)} e^{-2\delta_1 T}.
\tag{5.46}
$$

Here the left hand side is the operator norm as an endomorphism of $L^2_m(K_1)$.

On the other hand, (5.9), (5.34) and (5.43) imply

$$
\|\overline{\partial}(\mathrm{Exp}(u, V)) - \mathcal{P}^{-1}\mathfrak{e}\|_{L^2_m(K_1)} \leq C_{m,(5.47)} e^{-\delta_1 T}.
\tag{5.47}
$$

(Note we regard $\mathfrak{e} = \mathfrak{e}^\rho_{1,T,(0)} \in L^2(K_1, u^*TX \otimes \Lambda^{0,1})$.) We put

$$
Q = \overline{\partial}(\mathrm{Exp}(u, V)) - \mathcal{P}^{-1}\mathfrak{e}.
$$

By (5.45), (5.46) and (5.47), we have

$$\begin{aligned}&\left\|\Pi^{\perp}_{\mathcal{E}_1(\mathrm{Exp}(u,V))}(Q) - \mathcal{P}^{-1}\Pi^{\perp}_{\mathcal{E}_1(u)}(\mathcal{P}Q) + \mathcal{P}^{-1}(D_u\mathcal{E}_1)(\mathcal{P}Q, V)\right\|_{L^2_m(K_1)}\\ &= \left\|\int_0^1 ds \int_0^s \frac{d^2}{dr^2}\left(\mathcal{P}\circ\Pi^{\perp}_{\mathcal{E}_1(\mathrm{Exp}(u,rV))}(Q)\right) dr\right\|_{L^2_m(K_1)}\\ &\le C_{m,(5.48)}e^{-3\delta_1 T}.\end{aligned} \tag{5.48}$$

By (5.34) we have $\|V\|_{L^2_{m+1}(\Sigma_i)} \le C_{2,m}e^{-\delta_1 T}$. Therefore using (5.47), (5.48) and

$$\left\|\mathcal{P}^{-1}(D_u\mathcal{E}_1)(\mathcal{P}Q, V)\right\|_{L^2_m(K_1)} \le C_{m,(5.49)}\,\|V\|_{L^2_m(K_1)}\,\|Q\|_{L^2_m(K_1)} \tag{5.49}$$

we have

$$\left\|\Pi^{\perp}_{\mathcal{E}_1(\mathrm{Exp}(u,V))}(Q) - \mathcal{P}^{-1}\Pi^{\perp}_{\mathcal{E}_1(u)}(\mathcal{P}Q)\right\|_{L^2_m(K_1)} \le C_{m,(5.50)}e^{-2\delta_1 T}. \tag{5.50}$$

Substituting $Q = \overline{\partial}(\mathrm{Exp}(u,V)) - \mathcal{P}^{-1}\mathfrak{e}$ into (5.50) back, we obtain

$$\begin{aligned}&\Big\|\Pi^{\perp}_{\mathcal{E}_1(\mathrm{Exp}(u,V))}\overline{\partial}(\mathrm{Exp}(u,V)) - \mathcal{P}^{-1}\Pi^{\perp}_{\mathcal{E}_1(u)}(\mathcal{P}\overline{\partial}(\mathrm{Exp}(u,V))\\ &\quad - \Pi^{\perp}_{\mathcal{E}_1(\mathrm{Exp}(u,V))}\left(\mathcal{P}^{-1}\mathfrak{e}\right) + \mathcal{P}^{-1}\Pi^{\perp}_{\mathcal{E}_1(u)}(\mathfrak{e})\Big\|_{L^2_m(K_1)} \le C_{m,(5.51)}e^{-2\delta_1 T}.\end{aligned} \tag{5.51}$$

Therefore applying (5.45) and (5.46) to the 3rd and 4th terms of the left hand side of (5.51), we obtain

$$\begin{aligned}&\Big\|\Pi^{\perp}_{\mathcal{E}_1(\mathrm{Exp}(u,V))}\overline{\partial}(\mathrm{Exp}(u,V)) - \mathcal{P}^{-1}\Pi^{\perp}_{\mathcal{E}_1(u)}(\mathcal{P}\overline{\partial}(\mathrm{Exp}(u,V))\\ &\qquad + \mathcal{P}^{-1}(D_u\mathcal{E}_1)\left(\mathfrak{e}, V\right)\Big\|_{L^2_m(K_1)} \le C_{m,(5.52)}e^{-2\delta_1 T}.\end{aligned} \tag{5.52}$$

We recall $u = \hat{u}^{\rho}_{1,T,(0)} = u_1^{\rho_1}$ on K_1^+. Therefore we derive

$$\overline{\partial}u + (D_u\overline{\partial})(V) - (D_u\mathcal{E}_1)(\mathfrak{e}, V) \in \mathcal{E}_1(u) \tag{5.53}$$

on K_1^+ from (5.20), (5.30) and Definition 5.5.

Then, (5.52) and (5.53) imply

$$\begin{aligned}&\Big\|\Pi^{\perp}_{\mathcal{E}_1(\mathrm{Exp}(u,V))}\overline{\partial}(\mathrm{Exp}(u,V)) - \mathcal{P}^{-1}\Pi^{\perp}_{\mathcal{E}_1(u)}\mathcal{P}\overline{\partial}(\mathrm{Exp}(u,V))\\ &\quad + \mathcal{P}^{-1}\Pi^{\perp}_{\mathcal{E}_1(u)}\overline{\partial}u + \mathcal{P}^{-1}\Pi^{\perp}_{\mathcal{E}_1(u)}(D_u\overline{\partial})(V)\Big\|_{L^2_m(K_1)}\\ &\qquad \le C_{m,(5.54)}e^{-2\delta_1 T}.\end{aligned} \tag{5.54}$$

Combined with (5.43) and (5.44), this implies

$$\begin{aligned}\left\|\Pi^{\perp}_{\mathcal{E}_1(\mathrm{Exp}(u,V))}\left(\overline{\partial}(\mathrm{Exp}(u,V))\right)\right\|_{L^2_m(K_1)} &\le C_{m,(5.55)}e^{-2\delta_1 T}\\ &\le C_{1,m}e^{-\delta_1 T}\epsilon(4)\mu/10,\end{aligned} \tag{5.55}$$

for $T > T_{m,\epsilon(4),(5.55)}$ if we choose $T_{m,\epsilon(4),(5.55)}$ so that it satisfies

$$e^{-\delta_1 T_{m,\epsilon(4),(5.55)}} < \frac{C_{1,m}}{10C_{m,(5.55)}}\epsilon(4)\mu. \tag{5.56}$$

This gain of decay rate by the order of $\mu > 0$ independent of T and m is one of the crucial elements in the iteration scheme. (We refer readers to (5.81) to see how this gain is used.) We use (5.47) and (5.55) to show:

LEMMA 5.20. *There exists $C_{m,(5.57)}$ such that:*

$$\|\Pi_{\mathcal{E}_1(\mathrm{Exp}(u,V)))}(\overline{\partial}(\mathrm{Exp}(u,V)) - \mathfrak{e}\|_{L^2_m(K_1)} \le C_{m,(5.57)} e^{-\delta_1 T}. \tag{5.57}$$

Here we regard $\mathfrak{e} = \mathfrak{e}^{\rho}_{1,T,(0)}$ as an element of $L^2_m(K_1; (\mathrm{Exp}_1(u,V)))^*TX \otimes \Lambda^{0,1})$.

PROOF. This is a consequence of (5.47), $\mathfrak{e} \in \mathcal{E}_1(\mathrm{Exp}(u,V))$ and certain estimate of parallel transport. We postpone the proof of this lemma till Appendix B. □

Then if we set

$$\mathfrak{e}^{\rho}_{1,T,(1)} = \Pi_{\mathcal{E}_1(\mathrm{Exp}(u,V))}\left(\overline{\partial}\,(\mathrm{Exp}(u,V)) - \mathfrak{e}^{\rho}_{1,T,(0)}\right) \tag{5.58}$$

(5.39) follows. (Recall $\mathfrak{e} = \mathfrak{e}^{\rho}_{1,T,(0)}$.)

The estimate of the left hand side of (5.38) on K_1 follows from (5.55). We have thus finished the estimate on K_1. The estimate on K_2 is the same.

Step 1-3-2: Estimate on the neck region

We next consider the domain $[-5T, -T-1] \times [0,1]$. Let $S \in [-5T, -T-1]$ and

$$\Sigma(S) = [S, S+1] \times [0,1].$$

We also put $v = \Delta p^{\rho}_{T,(1)}$.

In the same way as the proof of Lemma 5.19 we can use (5.34) to prove

$$\begin{aligned}
&\left\| \int_0^1 ds \int_0^s \left(\frac{\partial}{\partial r}\right)^2 (\mathcal{P}\overline{\partial}(\mathrm{Exp}(u, rV))\, dr \right\|_{L^2_m(\Sigma(S))} \\
&\le C_{m,(5.59)} \|V\|^2_{L^2_{m+1}(\Sigma(S))} \\
&\le C_{m,(5.59)} \left(\|V - v^{\mathrm{Pal}}\|^2_{L^2_{m+1}(\Sigma(S))} + \|v^{\mathrm{Pal}}\|^2_{L^2_{m+1}(\Sigma(S))}\right) \\
&\le C'_{m,(5.59)} e^{-2\delta_1 T}.
\end{aligned} \tag{5.59}$$

(See Appendix A.) Here to prove the inequality of the last line we also use the boundedness of the domain $\Sigma(S)$ and an equality $\|v^{\mathrm{Pal}}\|_{L^2_{m+1}(\Sigma(S))} \le C\|v\|$, which follows from the definition (3.7) of v^{Pal}.

We remark that the domain $\Sigma(S)$ is disjoint from the supports of elements of $\mathcal{E}_i$. Therefore (5.43), (5.59) together with (5.30) imply

$$\|\overline{\partial}(\mathrm{Exp}(u,V))\|_{L^2_m(\Sigma(S))} \le C_{m,(5.60)} e^{-2\delta_1 T}. \tag{5.60}$$

Note $e_{T,\delta} \le 10 e^{5T\delta} \le 10 e^{T\delta_1/2}$ by (3.11). Hence (5.60) implies

$$\begin{aligned}
&\|\overline{\partial}(\mathrm{Exp}(u,V))\|^2_{L^2_{m,\delta}([-5T,-T-1]\times[0,1])} \\
&\le \sum_{S \in ([-5T-1,-T-2]\cap\mathbb{Z})\cup\{-T-2\}} \|\overline{\partial}(\mathrm{Exp}(u,V))\|^2_{L^2_{m,\delta}(\Sigma(S))} \\
&\le C_{m,(5.61)} T e^{-3\delta_1 T} \le \left(C_{1,m} e^{-\delta_1 T} \epsilon(4)\mu/10\right)^2
\end{aligned} \tag{5.61}$$

if $T > T_{m,\epsilon(4),(5.61)}$.

The inequality (5.61) is the estimate of the left hand side of (5.38) on

$$[-5T, -T-1] \times [0,1].$$

The estimate on $[T+1, 5T] \times [0,1]$ is the same. (We use the notations given in (5.41) up to here.)

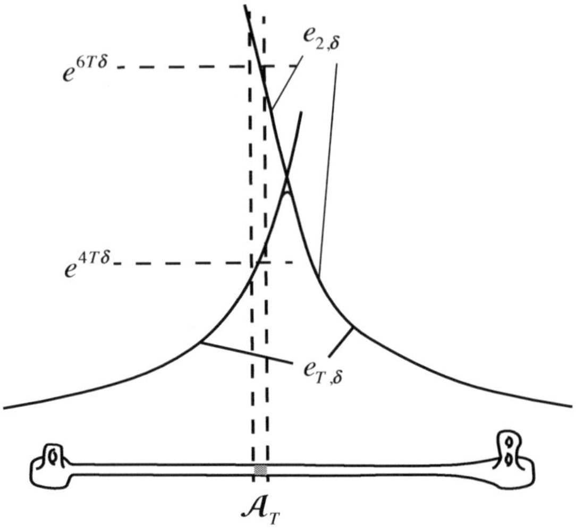

FIGURE 5.1. Drop of weight.

Now we do estimate $\overline{\partial} u^{\rho}_{T,(1)}$ on $[-T+1, T-1] \times [0,1]$. The inequality

$$\left\| \overline{\partial} u^{\rho}_{T,(1)} \right\|_{L^2_{m,\delta}([-T+1,T-1]\times[0,1]\subset\Sigma_T)} < C_{1,m}\mu\epsilon(4)e^{-\delta_1 T}/10$$

is also a consequence of the fact that (5.8) is the linearized equation of (3.19) and of the estimate (5.34). In fact, since the bump functions $\chi_{\mathcal{B}}^{\leftarrow}$ and $\chi_{\mathcal{A}}^{\rightarrow}$ are $\equiv 1$ there the proof is the same as the proof of (5.61).

On $\mathcal{A}_T$, by definition of $u^{\rho}_{T,(1)}$, we have

$$\overline{\partial} u^{\rho}_{T,(1)} = \overline{\partial}\left(\operatorname{Exp}\left(u^{\rho}_{T,(0)}, \chi_{\mathcal{A}}^{\rightarrow}(V^{\rho}_{T,2,(1)} - (\Delta p^{\rho}_{T,(1)})^{\mathrm{Pal}}) + V^{\rho}_{T,1,(1)}\right)\right). \tag{5.62}$$

Note

$$\begin{aligned}
&\left\| \chi_{\mathcal{A}}^{\rightarrow}(V^{\rho}_{T,2,(1)} - (\Delta p^{\rho}_{T,(1)})^{\mathrm{Pal}}) \right\|_{L^2_{m+1}(\mathcal{A}_T)} \\
&\le C_{m,(5.63)} e^{-6T\delta} \left\| V^{\rho}_{T,2,(1)} - (\Delta p^{\rho}_{T,(1)})^{\mathrm{Pal}} \right\|_{L^2_{m+1,\delta}(\mathcal{A}_T\subset\Sigma_2)} \\
&\le C'_{m,(5.63)} e^{-6\delta T - \delta_1 T}.
\end{aligned} \tag{5.63}$$

The first inequality follows from the fact that the weight function $e_{2,\delta}$ is around $e^{6T\delta}$ on $\mathcal{A}_T$. The second inequality follows from (5.34). On the other hand the weight function $e_{T,\delta}$ is around $e^{4\delta T}$ at $\mathcal{A}_T$. See Figure 5.1. Therefore

$$\begin{aligned}
&\left| \|\overline{\partial} u^{\rho}_{T,(1)}\|_{L^2_{m,\delta}(\mathcal{A}_T\subset\Sigma_T)} - \|\overline{\partial}(\operatorname{Exp}(u^{\rho}_{T,(0)}, V^{\rho}_{T,1,(1)})\|_{L^2_{m,\delta}(\mathcal{A}_T\subset\Sigma_T)} \right| \\
&\le C_{m,(5.64)} \|\chi_{\mathcal{A}}^{\rightarrow}(V^{\rho}_{T,2,(1)} - (\Delta p^{\rho}_{T,(1)})^{\mathrm{Pal}})\|_{L^2_{m+1,\delta}(\mathcal{A}_T\subset\Sigma_T)} \\
&\le C'_{m,(5.64)} e^{-2\delta T - \delta_1 T}.
\end{aligned} \tag{5.64}$$

See Appendix A for the proof of the first inequality.

REMARK 5.21. This drop of the weight is the main part of the idea. It was used in [**FOOO1**, page 414] (and also in [**Fu**, (8.7.2)]).

By Definition 5.5, $\mathrm{Err}^{\rho}_{2,T,(0)} = 0$ on $\mathcal{A}_T$. Using this in the same way as we did to obtain (5.60) we derive

$$\|\overline{\partial}(\mathrm{Exp}(u^{\rho}_{T,(0)}, V^{\rho}_{T,1,(1)})\|_{L^2_{m,\delta}(\mathcal{A}_T \subset \Sigma_T)} \leq C_{1,m} e^{-\delta_1 T}\epsilon(4)\mu/20 \tag{5.65}$$

for $T > T_{m,\epsilon(4),(5.66)}$.

Therefore combining (5.62)-(5.65), we can find a constant $T_{m,\epsilon(4),(5.66)}$ such that

$$\|\overline{\partial} u^{\rho}_{T,(1)}\|_{L^2_{m,\delta}(\mathcal{A}_T \subset \Sigma_T)} < C_{1,m}\mu\epsilon(4)e^{-\delta_1 T}/10 \tag{5.66}$$

for all $T > T_{m,\epsilon(4),(5.66)}$.

The estimate on $\mathcal{B}_T$ is similar and so omitted. The proof of Proposition 5.17 is now complete. □

Step 1-4 (Separating error terms into two parts): We split the error (5.40) into the two parts as follows.

DEFINITION 5.22. We put

$$\mathrm{Err}^{\rho}_{1,T,(1)} = \chi^{\leftarrow}_{\mathcal{X}} \mathrm{Err}^{\rho}_{T,(1)} = \chi^{\leftarrow}_{\mathcal{X}}(\overline{\partial} u^{\rho}_{T,(1)} - (\mathfrak{e}^{\rho}_{1,T,(0)} + \mathfrak{e}^{\rho}_{1,T,(1)})),$$
$$\mathrm{Err}^{\rho}_{2,T,(1)} = \chi^{\rightarrow}_{\mathcal{X}} \mathrm{Err}^{\rho}_{T,(1)} = \chi^{\rightarrow}_{\mathcal{X}}(\overline{\partial} u^{\rho}_{T,(1)} - (\mathfrak{e}^{\rho}_{2,T,(0)} + \mathfrak{e}^{\rho}_{2,T,(1)})).$$

We regard them as elements of the weighted Sobolev spaces

$$L^2_{m,\delta}(\Sigma_1; (\hat{u}^{\rho}_{1,T,(1)})^* TX \otimes \Lambda^{0,1})$$

and $L^2_{m,\delta}(\Sigma_2; (\hat{u}^{\rho}_{2,T,(1)})^* TX \otimes \Lambda^{0,1})$ respectively, by extending them to be 0 outside the support of χ.

We put

$$p^{\rho}_{T,(1)} = \mathrm{Exp}(p^{\rho}, \Delta p^{\rho}_{T,(1)}). \tag{5.67}$$

5.3. The general κ-th iteration with $\kappa \geq 1$

We then come back to the Step κ-1 for $\kappa = 1$ to establish the following and continue our inductive steps for $\kappa \geq 1$.

$$\left\|(V^{\rho}_{T,i,(\kappa)}, \Delta p^{\rho}_{T,(\kappa)})\right\|_{W^2_{m+1,\delta}(\Sigma_i)} < C_{2,m}\mu^{\kappa-1}e^{-\delta_1 T}, \tag{5.68}$$

$$\left\|\mathrm{E}(u^{\rho}_{T,(0)}, u^{\rho}_{T,(\kappa)})\right\|_{W^2_{m+1,\delta}(\Sigma_T)} < C_{4,m}(2-\mu^{\kappa})e^{-\delta_1 T}, \tag{5.69}$$

$$\left\|\mathrm{Err}^{\rho}_{i,T,(\kappa)}\right\|_{L^2_{m,\delta}(\Sigma_i)} < C_{1,m}\epsilon(5)\mu^{\kappa}e^{-\delta_1 T}, \tag{5.70}$$

$$\left\|\mathfrak{e}^{\rho}_{i,T,(\kappa)}\right\|_{L^2_m(K_i)} < C_{3,m}\mu^{\kappa-1}e^{-\delta_1 T}, \quad \text{for } \kappa \geq 1. \tag{5.71}$$

Note we also have

$$\overline{\partial} u^{\rho}_{T,(\kappa)} - \mathrm{Err}^{\rho}_{1,T,(\kappa)} - \mathrm{Err}^{\rho}_{2,T,(\kappa)} = \sum_{i=1}^{2}\sum_{j=0}^{\kappa} \mathfrak{e}^{\rho}_{i,T,(j)}. \tag{5.72}$$

See Definitions 5.22 and 5.35.

Formulas (5.68)-(5.71) are proved by induction on κ. Since various constants appear during the inductive proof of (5.68)-(5.71) it is important to clarify the dependence between them. We state the inductive scheme of the proof of Formulas (5.68)-(5.71) clarifying the dependence of the various constants as Proposition 5.22 below.

Note we have already chosen the constants $C_{1,m}, C_{2,m}$, and $C_{3,m}$ in (5.9), (5.34), and (5.39), respectively.

PROPOSITION 5.23.

(1) *For any* $\epsilon(5) > 0$ *and* $C_{4,m}$ *there exists* $T_{3,m,\epsilon(5)} > 0$ *such that if* $T > T_{3,m,\epsilon(5)}$ *then* (5.68), (5.69) *for* $\kappa \le \kappa_0$ *and* (5.70) *and* (5.71) *for* $\kappa \le \kappa_0 - 1$ *imply* (5.70) *and* (5.71) *for* $\kappa = \kappa_0$.
(2) *For any* $C_{4,m}$ *there exists* $\epsilon_{3,m} > 0$ *independent of* T, κ_0 *such that if* $\epsilon(5) < \epsilon_{3,m}$*, then* (5.70) *and* (5.71) *for* $\kappa \le \kappa_0 - 1$ *imply* (5.68) *for* $\kappa = \kappa_0$.
(3) *We can choose* $C_{4,m}$ *such that the following holds. There exists* $T_{4,m} > 0$ *such that if* (5.68) *holds for* $\kappa \le \kappa_0$, (5.69) *holds for* $\kappa \le \kappa_0 - 1$ *and* $T > T_{4,m}$ *then* (5.69) *holds for* $\kappa = \kappa_0$.

See Figure 5.2.

REMARK 5.24.

(1) The constant $T_{3,m,\epsilon(5)}$ in Proposition 5.23 (1) may depend on $C_{1,m}$, $C_{2,m}$, $C_{3,m}$, $C_{4,m}$.
(2) The constant $\epsilon_{3,m}$ in Proposition 5.23 (2) may depend on $C_{1,m}$, $C_{2,m}$, $C_{3,m}$, $C_{4,m}$.
(3) The constant $T_{4,m}$ in Proposition 5.23 (3) may depend on $C_{1,m}$, $C_{2,m}$, $C_{3,m}$, $C_{4,m}$.
(4) In Proposition 5.23 (3) the map $u^{\rho}_{T,(\kappa)}$ $u^{\rho}_{T,(\kappa)}$ for $\kappa \le \kappa_0$, is defined from $u^{\rho}_{T,(\kappa-1)}$ and $(V^{\rho}_{T,i,(\kappa)}, \Delta p^{\rho}_{T,(\kappa)})$ as in Definition 5.33 and (5.69).
(5) It is important that the constants appearing in Proposition 5.23 are independent of κ_0.

We prove Formulas (5.68)-(5.71) using Proposition 5.23 as follows. We first fix $C_{4,m}$ so that Proposition 5.23 (3) holds. Then $T_{3,m,\epsilon(5)}$ (depending on $\epsilon(5)$) is determined by Proposition 5.23 (1). $\epsilon_{3,m}$ is determined by Proposition 5.23 (2). $T_{4,m}$ is determined by Proposition 5.23 (3). Formulas (5.68)-(5.71) hold for $T \ge \max\{T_{3,m,\epsilon(5)}, T_{4,m}\}$ by Proposition 5.23. We may take $\epsilon(5) = \epsilon_{3,m}/2$ and so $T_{3,m,\epsilon(5)} = T_{3,m,\epsilon_{3,m}/2}$ is determined.

PROOF OF PROPOSITION 5.23. The rest of this chapter will be occupied with the proof of this proposition by describing each of Steps κ-1,...,κ-4. For given κ_0 the Step κ_0-1 corresponds to Proposition 5.23 (2). Steps κ_0-2,κ_0-3,κ_0-4 correspond to Proposition 5.23 (1). The proof of Proposition 5.23 (3) is straightforward. We provide its detail in Appendix E together with the versions involving T (and ρ) derivatives. See Proposition 6.9 (3).

Since Steps κ-1,...,κ-4 are largely repetitions of the same kind of tedious but straightforward estimates as in $\kappa = 1$, we will only state the main definitions and statements required to perform the inductive schemes that will be needed for the exponential estimate of the T-derivatives in the next section, and leave the details of the relevant estimates to Appendix C.

REMARK 5.25. Various constants $C_{m,(*.*)}$ and $T_{m,\epsilon(5),(*.*)}$ etc. will appear in the course of the proof of Proposition 5.23 (including the appendices quoted there.)

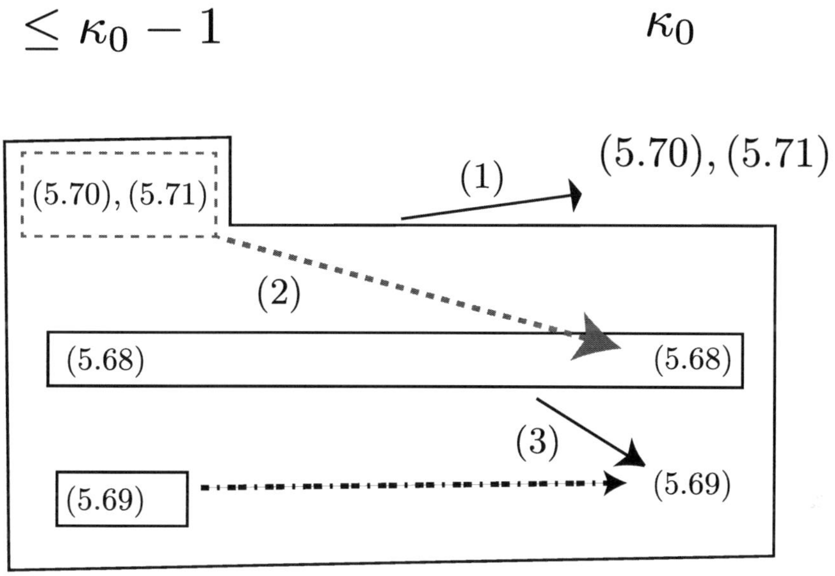

FIGURE 5.2. Logical scheme in Proposition 5.23

Those constants depend on m as well as the numbers explicitly appearing in the subscript and the previously given constants $C_{1,m}$-$C_{4,m}$. (Especially they depend on the $W_{m+1,\delta}$ norm of $u^{\rho}_{T,(\kappa)}$ which is estimated by (5.69).) The important point is that they are independent of κ and T.

Therefore, even though we have infinitely many steps $\kappa = 0, 1, 2, \dots$ to work out, we need to make choices of those constants only finitely many times.

Step κ-1: In this step we prove Proposition 5.23 (2). We first cut-off $u^{\rho}_{T,(\kappa-1)}$ and extend it by a constant to obtain the maps $\hat{u}^{\rho}_{i,T,(\kappa-1)} : (\Sigma_i, \partial\Sigma_i) \to (X, L)$ $(i = 1, 2)$ as follows: We first define $p^{\rho}_{T,(\kappa-1)}$ inductively by

$$p^{\rho}_{T,(\kappa-1)} = \mathrm{Exp}(p^{\rho}_{T,(\kappa-2)}, \Delta p^{\rho}_{T,(\kappa-1)}), \tag{5.73}$$

starting from (5.67). Then we put

$$
(5.74)\quad
\begin{aligned}
&\hat{u}^\rho_{1,T,(\kappa-1)}(z)\\
&=\begin{cases}
\mathrm{Exp}\left(p^\rho_{T,(\kappa-1)}, \chi^{\leftarrow}_{\mathcal{B}}(\tau-T,t)\mathrm{E}(p^\rho_{T,(\kappa-1)}, u^\rho_{T,(\kappa-1)}(\tau,t))\right) & \\
 & \text{if } z=(\tau,t)\in[-5T,5T]\times[0,1]\\
u^\rho_{T,(\kappa-1)}(z) & \text{if } z\in K_1\\
p^\rho_{T,(\kappa-1)} & \text{if } z\in[5T,\infty)\times[0,1].
\end{cases}\\
&\hat{u}^\rho_{2,T,(\kappa-1)}(z)\\
&=\begin{cases}
\mathrm{Exp}\left(p^\rho_{T,(\kappa-1)}, \chi^{\rightarrow}_{\mathcal{A}}(\tau+T,t)\mathrm{E}(p^\rho_{T,(\kappa-1)}, u^\rho_{T,(\kappa-1)}(\tau,t))\right) & \\
 & \text{if } z=(\tau,t)\in[-5T,5T]\times[0,1]\\
u^\rho_{T,(\kappa-1)}(z) & \text{if } z\in K_2\\
p^\rho_{T,(\kappa-1)} & \text{if } z\in(-\infty,-5T]\times[0,1].
\end{cases}
\end{aligned}
$$

We have the following estimates of $\hat{u}^\rho_{i,T,(\kappa-1)}$ on the neck region.

LEMMA 5.26.

$$
(5.75)\quad
\begin{aligned}
\left\|\mathrm{E}(p^\rho_0, \hat{u}^\rho_{1,T,(\kappa-1)})\right\|_{L^2_{m+1}([T,9T)\times[0,1])} &\le C_{m,(5.75)}e^{-\delta_1 T},\\
\left\|\mathrm{E}(p^\rho_0, \hat{u}^\rho_{2,T,(\kappa-1)})\right\|_{L^2_{m+1}((-9T,-T]\times[0,1])} &\le C_{m,(5.75)}e^{-\delta_1 T}.
\end{aligned}
$$

We remark that in the left hand side of (5.75) we take the L^2_{m+1} norm *without* weight.

PROOF. For $\kappa-1=0$ this is a consequence of Condition 3.1 and Lemma 2.5. Then using (5.68) we can prove the lemma by induction on κ. Its version with T and ρ derivatives included, is (6.11) proven in Appendix E. □

Now we consider the linearization of (3.19) at $\hat{u}^\rho_{i,T,(\kappa-1)}$, which is

$$
(5.76)\quad
\begin{aligned}
D_{\hat{u}^\rho_{i,T,(\kappa-1)}}\overline{\partial} : W^2_{m+1,\delta}((\Sigma_i,\partial\Sigma_i);(\hat{u}^\rho_{i,T,(\kappa-1)})^*TX, (\hat{u}^\rho_{i,T,(\kappa-1)})^*TL)&\\
\to L^2_{m,\delta}(\Sigma_i;(\hat{u}^\rho_{i,T,(\kappa-1)})^*TX\otimes\Lambda^{0,1}).&
\end{aligned}
$$

We denote

$$
(5.77)\quad (\mathfrak{se})^\rho_{i,T,(\kappa-1)} = \sum_{a=0}^{\kappa-1}\mathfrak{e}^\rho_{i,T,(a)}.
$$

Similarly as for the operator $D^{\mathrm{app},(0)}_{\hat{u}^\rho_{i,T,(0)}}$ in (5.20), we define the approximate linearization of (3.19)

$$
(5.78)\quad D^{\mathrm{app},(\kappa-1)}_{\hat{u}^\rho_{i,T,(\kappa-1)}} := D_{\hat{u}^\rho_{i,T,(\kappa-1)}}\overline{\partial} - (D_{\hat{u}^\rho_{i,T,(\kappa-1)}}\mathcal{E}_i)((\mathfrak{se})^\rho_{i,T,(\kappa-1)},\cdot)
$$

replacing $\overline{\partial}\hat{u}^\rho_{i,T,(\kappa-1)}$ by $(\mathfrak{se})^\rho_{i,T,(\kappa-1)}$ in the expression of the linearization operator

$$
(D_{\hat{u}^\rho_{i,T,(\kappa-1)}}\overline{\partial})^\perp = D_{\hat{u}^\rho_{i,T,(\kappa-1)}}\overline{\partial} - (D_{\hat{u}^\rho_{i,T,(\kappa-1)}}\mathcal{E}_i)(\overline{\partial}\hat{u}^\rho_{i,T,(\kappa-1)},\cdot).
$$

LEMMA 5.27. *Denote* $\mathcal{E}_i = \mathcal{E}_i(\hat{u}^\rho_{i,T,(\kappa-1)})$. *There exists* $T_{m,(5.79)}$ *such that if* $T > T_{m,(5.79)}$ *then we have*

$$\operatorname{Im}(D^{\mathrm{app},(\kappa-1)}_{\hat{u}^\rho_{i,T,(\kappa-1)}}) + \mathcal{E}_i = L^2_{m,\delta}(\Sigma_i; (\hat{u}^\rho_{i,T,(\kappa-1)})^*TX \otimes \Lambda^{0,1}). \tag{5.79}$$

Moreover the map

$$\mathrm{Dev}_{1,\infty} - \mathrm{Dev}_{2,\infty} : (D^{\mathrm{app},(\kappa-1)}_{\hat{u}^\rho_{1,T,(\kappa-1)}})^{-1}(\mathcal{E}_1) \oplus (D^{\mathrm{app},(\kappa-1)}_{\hat{u}^\rho_{2,T,(\kappa-1)}})^{-1}(\mathcal{E}_2) \to T_{p^\rho_{T,(\kappa-1)}} L \tag{5.80}$$

is surjective.

PROOF. Using the inequality (5.71), we estimate

$$\left\| \mathfrak{e}^\rho_{i,T,(0)} - \sum_{a=0}^{\kappa-1} \mathfrak{e}^\rho_{i,T,(a)} \right\|_{L^2_m(K_i)} \leq \sum_{a=1}^{\kappa-1} \|\mathfrak{e}^\rho_{i,T,(a)}\|_{L^2_m(K_i)} < C_{3,m} \frac{e^{-\delta_1 T}}{1-\mu}. \tag{5.81}$$

This in particular implies that

$$\begin{aligned} Y \mapsto &(D_{\hat{u}^\rho_{1,T,(\kappa-1)}} \mathcal{E}_1)((\mathfrak{se})^\rho_{i,T,(\kappa-1)}, \mathrm{Pal}_{u_1}^{\hat{u}^\rho_{1,T,(\kappa-1)}}(Y)) \\ &- \mathrm{Pal}_{\hat{u}^\rho_{1,T,(0)}}^{\hat{u}^\rho_{1,T,(\kappa-1)}}((D_{\hat{u}^\rho_{1,T,(0)}} \mathcal{E}_1)(\mathfrak{e}^\rho_{1,T,(0)}, \mathrm{Pal}_{u_1}^{\hat{u}^\rho_{1,T,(0)}}(Y))) \end{aligned} \tag{5.82}$$

is small in the operator norm. We prove this fact in Appendix D. Then this smallness implies that the operator norm of the difference operator

$$D^{\mathrm{app},(\kappa-1)}_{\hat{u}^\rho_{1,T,(\kappa-1)}} \circ \mathrm{Pal}_{u^\rho_1}^{\hat{u}^\rho_{1,T,(\kappa-1)}} - \mathrm{Pal}_{\hat{u}^\rho_{1,T,(0)}}^{\hat{u}^\rho_{1,T,(\kappa-1)}} \circ D^{\mathrm{app},(0)}_{\hat{u}^\rho_{1,T,(0)}} \circ \mathrm{Pal}_{u^\rho_1}^{\hat{u}^\rho_{1,T,(0)}}$$

is small. This can be easily seen by rewriting the value of the difference operator as

$$\begin{aligned} &D^{\mathrm{app},(\kappa-1)}_{\hat{u}^\rho_{1,T,(\kappa-1)}}(\mathrm{Pal}_{u^\rho_1}^{\hat{u}^\rho_{1,T,(\kappa-1)}}(Y)) - \mathrm{Pal}_{\hat{u}^\rho_{1,T,(0)}}^{\hat{u}^\rho_{1,T,(\kappa-1)}}(D^{\mathrm{app},(0)}_{\hat{u}_{1,T,(0)}}(\mathrm{Pal}_{u^\rho_1}^{\hat{u}^\rho_{1,T,(0)}}(Y))) \\ &= \Big((D_{\hat{u}^\rho_{1,T,(\kappa-1)}} \overline{\partial})(\mathrm{Pal}_{u^\rho_1}^{\hat{u}^\rho_{1,T,(\kappa-1)}}(Y)) - \mathrm{Pal}_{\hat{u}^\rho_{1,T,(0)}}^{\hat{u}^\rho_{1,T,(\kappa-1)}}((D_{\hat{u}^\rho_{i,T,(0)}} \overline{\partial})(\mathrm{Pal}_{u^\rho_1}^{\hat{u}^\rho_{1,T,(0)}}(Y)))\Big) \\ &\quad - \Big((D_{\hat{u}^\rho_{1,T,(\kappa-1)}} \mathcal{E}_1)((\mathfrak{se})^\rho_{i,T,(\kappa-1)}, \mathrm{Pal}_{u^\rho_1}^{\hat{u}^\rho_{1,T,(\kappa-1)}} Y) \\ &\qquad - \mathrm{Pal}_{\hat{u}^\rho_{1,T,(0)}}^{\hat{u}^\rho_{1,T,(\kappa-1)}}(D_{\hat{u}^\rho_{1,T,(0)}} \mathcal{E}_1)(\mathfrak{e}^\rho_{1,T,(0)}, \mathrm{Pal}_{u^\rho_1}^{\hat{u}^\rho_{1,T,(0)}}(Y))\Big). \end{aligned}$$

The first summand of the right hand side is clearly small by the smallness of $L^2_{m+1,\delta}$-norm of $\mathrm{E}(\hat{u}^\rho_{i,T,(\kappa-1)}, \hat{u}^\rho_{i,T,(0)})$ arising from the induction hypothesis (5.69). Then the lemma follows by combining Lemma 5.7 and the above mentioned smallness. □

Note that Remark 5.8 still applies to Lemma 5.27.

DEFINITION 5.28. We define a linear map

$$\begin{aligned} \Phi_{i;(\kappa-1)}(\rho, T) :& W^2_{m+1,\delta}((\Sigma_i, \partial\Sigma_i); u_i^*TX, u_i^*TL) \\ &\to W^2_{m+1,\delta}((\Sigma_i, \partial\Sigma_i); (\hat{u}^\rho_{i,T,(\kappa-1)})^*TX, (\hat{u}^\rho_{i,T,(\kappa-1)})^*TL), \end{aligned}$$

as the closure of the map $\mathrm{Pal}_{u_i}^{\hat{u}^\rho_{i,T,(\kappa-1)}}$ given in (2.7).

LEMMA 5.29.

(1) $\Phi_{i;(\kappa-1)}(\rho, T)$ *sends* (5.26) *to*

$$W^2_{m+1,\delta}((\Sigma_i, \partial\Sigma_i); (\hat{u}^\rho_{i,T,(\kappa-1)})^*TX, (\hat{u}^\rho_{i,T,(\kappa-1)})^*TL). \tag{5.83}$$

(2) *The map* $(\Phi_{1;(\kappa-1)}(\rho,T),\Phi_{2;(\kappa-1)}(\rho,T))$ *sends the subspace* (5.24) *to the subspace*

$$\begin{aligned}&\mathrm{Ker}(Dev_{1,\infty}-Dev_{2,\infty})\\ &\cap\bigoplus_{i=1}^{2}W^2_{m+1,\delta}((\Sigma_i,\partial\Sigma_i);(\hat u^{\rho}_{i,T,(\kappa-1)})^*TX,(\hat u^{\rho}_{i,T,(\kappa-1)})^*TL).\end{aligned}\tag{5.84}$$

The proof is easy and is omitted.

DEFINITION 5.30. We denote by $\mathfrak{H}_{(\kappa-1)}(\mathcal{E}_1,\mathcal{E}_2;\rho,T)$ the image of the subspace $\mathfrak{H}(\mathcal{E}_1,\mathcal{E}_2)$ (See Definition 5.9.) under the map $(\Phi_{1;(\kappa-1)}(\rho,T),\Phi_{2;(\kappa-1)}(\rho,T))$. It is a subspace of (5.84).

DEFINITION 5.31. We define $(V^{\rho}_{T,1,(\kappa)},V^{\rho}_{T,2,(\kappa)},\Delta p^{\rho}_{T,(\kappa)})$ by

$$D^{\mathrm{app},(\kappa-1)}_{\hat u^{\rho}_{i,T,(\kappa-1)}}(V^{\rho}_{T,i,(\kappa)})+\mathrm{Err}^{\rho}_{i,T,(\kappa-1)}\in\mathcal{E}_i(\hat u^{\rho}_{i,T,(\kappa-1)})\tag{5.85}$$

and

$$((V^{\rho}_{T,1,(\kappa)},\Delta p^{\rho}_{T,(\kappa)}),(V^{\rho}_{T,2,(\kappa)},\Delta p^{\rho}_{T,(\kappa)}))\in\mathfrak{H}_{(\kappa-1)}(\mathcal{E}_1,\mathcal{E}_2;\rho,T).\tag{5.86}$$

Lemma 5.27 implies that such $(V^{\rho}_{T,1,(\kappa)},V^{\rho}_{T,2,(\kappa)},\Delta p^{\rho}_{T,(\kappa)})$ exists and is unique if $T>T_{m,(5.79)}$.

LEMMA 5.32. *There exist* $T_{m,(5.87)}$ *and* $\epsilon_{3,m}>0$ *such that if* $\epsilon(5)<\epsilon_{3,m}$, *then* (5.70) *for* $\kappa-1$ *implies that the next inequality for* $T>T_{m,(5.87)}$.

$$\|(V_{T,i,(\kappa)},\Delta p^{\rho}_{T,(\kappa)})\|_{W^2_{m+1,\delta}(\Sigma_i)}\le C_{2,m}\mu^{\kappa-1}e^{-\delta_1 T}.\tag{5.87}$$

PROOF. We take $T_{m,(5.87)}$ so large that the inverse of

$$\Pi^{\perp}_{\mathcal{E}_i(\hat u^{\rho}_{i,T,(\kappa-1)})}\circ D^{\mathrm{app},(\kappa-1)}_{\hat u^{\rho}_{i,T,(\kappa-1)}}$$

is uniformly bounded and choose $\epsilon_{3,m}$ so small that $\epsilon_{3,m}^{-1}$ is much larger than the norm of the above mentioned inverse. Then (5.87) follows from uniform boundedness of the above mentioned inverse together with (5.70) for $\kappa-1$. □

This lemma implies the inequality (5.68). We have finished the proof of Proposition 5.23 (2).

Step κ-2: We start the proof of Proposition 5.23 (1).

We use $(V^{\rho}_{T,1,(\kappa)},V^{\rho}_{T,2,(\kappa)},\Delta p^{\rho}_{T,(\kappa)})$ to find an approximate solution $u^{\rho}_{T,(\kappa)}$ of the next level. We remark that $(V^{\rho}_{T,1,(\kappa)},V^{\rho}_{T,2,(\kappa)},\Delta p^{\rho}_{T,(\kappa)})$ is the counterpart of v_κ given by $v_\kappa=\frac{f(x_\kappa)}{f'(x_\kappa)}$ and the next definition corresponds to the next iteration $x_{\kappa+1}=x_\kappa+v_\kappa$ in Newton's scheme.

DEFINITION 5.33. We define $u^{\rho}_{T,(\kappa)}(z)$ as follows.

(1) If $z\in K_1$, we put

$$u^{\rho}_{T,(\kappa)}(z)=\mathrm{Exp}(\hat u^{\rho}_{1,T,(\kappa-1)}(z),V^{\rho}_{T,1,(\kappa)}(z)).\tag{5.88}$$

(2) If $z\in K_2$, we put

$$u^{\rho}_{T,(\kappa)}(z)=\mathrm{Exp}(\hat u^{\rho}_{2,T,(\kappa-1)}(z),V^{\rho}_{T,2,(\kappa)}(z)).\tag{5.89}$$

(3) If $z = (\tau, t) \in [-5T, 5T] \times [0, 1]$, we put

$$u^{\rho}_{T,(\kappa)}(\tau,t) = \operatorname{Exp}\Big(u^{\rho}_{T,(\kappa-1)}(\tau,t), \chi^{\leftarrow}_{\mathcal{B}}(\tau,t)(V^{\rho}_{T,1,(\kappa)}(\tau,t) - (\Delta p^{\rho}_{T,(\kappa)})^{\mathrm{Pal}}) + \chi^{\rightarrow}_{\mathcal{A}}(\tau,t)(V^{\rho}_{T,2,(\kappa)}(\tau,t) - (\Delta p^{\rho}_{T,(\kappa)})^{\mathrm{Pal}}) + (\Delta p^{\rho}_{T,(\kappa)})^{\mathrm{Pal}}\Big). \tag{5.90}$$

We note that $\hat{u}^{\rho}_{1,T,(\kappa-1)}(z) = u^{\rho}_{T,(\kappa-1)}(z)$ on K_1 and $\hat{u}^{\rho}_{2,T,(\kappa-1)}(z) = u^{\rho}_{T,(\kappa-1)}(z)$ on K_2.

Step κ-3: The proof of the following proposition is largely a duplication of that of Proposition 5.17 plus (5.61) and so its details will be postponed till Appendix C.

PROPOSITION 5.34. *For any $\epsilon(5) > 0$, there exists $T_{m,\epsilon(5),(5.91)} > 0$ with the following properties. If $T > T_{m,\epsilon(5),(5.91)}$ we can define $\mathfrak{e}^{\rho}_{i,T,(\kappa)} \in \mathcal{E}_i$ that satisfies*

$$\left\| \overline{\partial} u^{\rho}_{T,(\kappa)} - \sum_{a=0}^{\kappa} \mathfrak{e}^{\rho}_{1,T,(a)} - \sum_{a=0}^{\kappa} \mathfrak{e}^{\rho}_{2,T,(a)} \right\|_{L^2_{m,\delta}(\Sigma_T)} < C_{1,m}\mu^{\kappa}\epsilon(5)e^{-\delta_1 T}, \tag{5.91}$$

and

$$\|\mathfrak{e}^{\rho}_{i,T,(\kappa)}\|_{L^2_m(K_i)} < C_{3,m}\mu^{\kappa-1}e^{-\delta_1 T}. \tag{5.92}$$

Note Remark 5.4 applies here.

Step κ-4: Similarly as before, we introduce the following definitions.

$$\mathrm{Err}^{\rho}_{T,(\kappa)} = \overline{\partial} u^{\rho}_{T,(\kappa)} - \sum_{a=0}^{\kappa} \mathfrak{e}^{\rho}_{1,T,(a)} - \sum_{a=0}^{\kappa} \mathfrak{e}^{\rho}_{2,T,(a)}. \tag{5.93}$$

DEFINITION 5.35. We put

$$\mathrm{Err}^{\rho}_{1,T,(\kappa)} = \chi^{\leftarrow}_{\mathcal{X}}(\mathrm{Err}^{\rho}_{T,(\kappa)}) = \chi^{\leftarrow}_{\mathcal{X}}\left(\overline{\partial} u^{\rho}_{T,(\kappa)} - \sum_{a=0}^{\kappa} \mathfrak{e}^{\rho}_{1,T,(a)}\right),$$

$$\mathrm{Err}^{\rho}_{2,T,(\kappa)} = \chi^{\rightarrow}_{\mathcal{X}}(\mathrm{Err}^{\rho}_{T,(\kappa)}) = \chi^{\rightarrow}_{\mathcal{X}}\left(\overline{\partial} u^{\rho}_{T,(\kappa)} - \sum_{a=0}^{\kappa} \mathfrak{e}^{\rho}_{2,T,(a)}\right).$$

We regard them as elements of the weighted Sobolev spaces

$$L^2_{m,\delta}(\Sigma_i; (\hat{u}^{\rho}_{i,T,(\kappa)})^*TX \otimes \Lambda^{0,1})$$

for $i = 1, 2$ respectively. (We extend them by 0 outside a compact set.)

We put $p^{\rho}_{T,(\kappa)} = \operatorname{Exp}(p^{\rho}_{T,(\kappa-1)}, \Delta p^{\rho}_{T,(\kappa)})$. See (5.73). Then Proposition 5.34 implies (5.70) and (5.71).

We have thus finished the proof of Proposition 5.23 (1).
The proof of Proposition 5.23 is complete. □

Now we are ready to produce the solution to the gluing problem. For each fixed m there exists T_m such that if $T > T_m$ then

$$\lim_{\kappa \to \infty} u^{\rho}_{T,(\kappa)}$$

converges in $L^2_{m+1,\delta}$ sense to a solution of (3.19) by Proposition 5.34. The limit map is automatically of C^{∞}-class by elliptic regularity. (5.70) and (5.72) imply that the limit solves the equation (3.19). We have thus constructed the gluing map used in Theorem 3.13.

CHAPTER 6

Exponential decay of T derivatives

We first state the result of this chapter which is also the main result of this paper. We recall that for T sufficiently large and $\rho = (\rho_1, \rho_2) \in V_1 \times_L V_2$ we have defined $u^{\rho}_{T,(\kappa)}$ for each $\kappa = 0, 1, 2, \ldots$. We denote its limit by

$$u^{\rho}_{T} = \lim_{\kappa \to \infty} u^{\rho}_{T,(\kappa)} : (\Sigma_T, \partial \Sigma_T) \to (X, L). \tag{6.1}$$

The main result of this chapter is an estimate of T and ρ derivatives of this map. We prepare some notations to state the result.

6.1. Statement of the main result

For $T > 0$, we defined the curve Σ_T as the union

$$\Sigma_T = K_1 \cup ([-5T, 5T]_{\tau} \times [0,1]) \cup K_2$$

by identifying $\partial K_1 \cong \{-5T\} \times [0,1]$ and $\partial K_2 \cong \{5T\} \times [0,1]$ where we denote by (τ, t) the coordinates on $[-5T, 5T]_{\tau} \times [0,1]$. See (3.1). Therefore we have the natural inclusion $K_i \subset \Sigma_T$.

We introduce new coordinates (τ', t) and (τ'', t) such that the relationship between the three coordinates (τ, t), (τ', t) and (τ'', t) are given by

$$\tau' = \tau + 5T \tag{6.2}$$

and

$$\tau'' = \tau - 5T. \tag{6.3}$$

See Figure 6.1. We may use either (τ', t) or (τ'', t) as the coordinates of $\Sigma_T \setminus (K_1 \cup K_2)$. Namely

$$\Sigma_T = K_1 \cup ([0, 10T]_{\tau'} \times [0,1]) \cup K_2 = K_1 \cup ([-10T, 0]_{\tau''} \times [0,1]) \cup K_2$$

REMARK 6.1. Here and hereafter we write $[0, 10T]_{\tau'}$, $[-10T, 0]_{\tau''}$, $[-5T, 5T]_{\tau}$ etc. to clarify the coordinate (τ', τ'', or τ) we use.

We can also use the (τ', t) coordinates on $\Sigma_1 \setminus K_1$ and the (τ'', t) coordinates on $\Sigma_2 \setminus K_2$ so that

$$\begin{aligned}
\Sigma_1 &= K_1 \cup ([-5T, \infty)_{\tau} \times [0,1]) = K_1 \cup ([0, \infty)_{\tau'} \times [0,1], \\
\Sigma_2 &= K_2 \cup (-\infty, 5T]_{\tau} \times [0,1]) = K_2 \cup (-\infty, 0]_{\tau''} \times [0,1].
\end{aligned}$$

REMARK 6.2. We remark that the (τ', t) coordinates of Σ_1 and the (τ'', t) coordinates of Σ_2 are independent of T but the (τ, t) coordinates of them are T dependent.

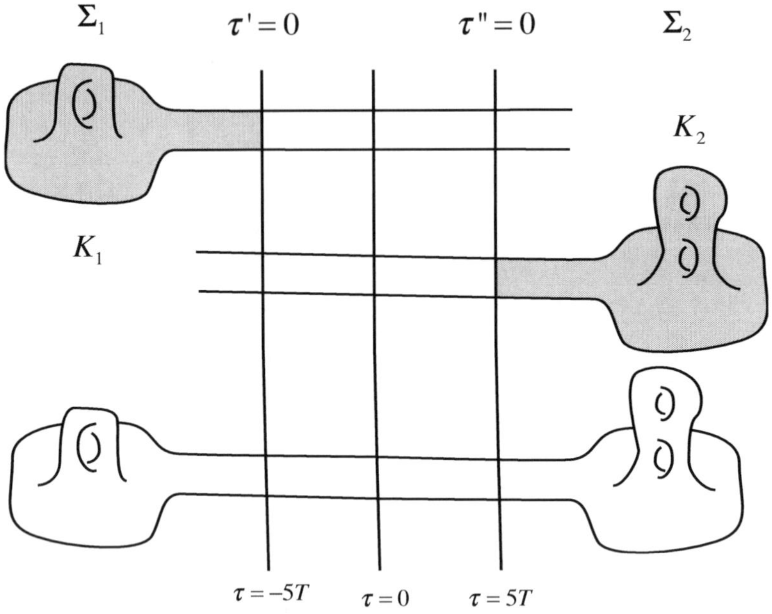

FIGURE 6.1. τ' and τ''.

We define

$$K_1^S = K_1 \cup ([0,S]_{\tau'} \times [0,1]),$$
$$K_2^S = ([-S,0]_{\tau''} \times [0,1]) \cup K_2 \tag{6.4}$$

for each positive constant $S > 0$. We have a natural embedding $K_1^S \hookrightarrow \Sigma_1$ (resp. $K_2^S \hookrightarrow \Sigma_2$) via the coordinates τ' (resp. via the coordinates τ'').

If $S > 0$, $T > \frac{S}{10}$, we have embeddings of $K_1^S \to \Sigma_T$ by setting $\tau = \tau' - 5T$ and $K_2^S \to \Sigma_T$ by $\tau = \tau'' + 5T$.

We then obtain a map

$$\mathrm{Glures}_{i,S} : [T_m, \infty) \times V_1 \times_L V_2 \to \mathrm{Map}_{L^2_{m+1}}((K_i^S, K_i^S \cap \partial\Sigma_i), (X, L))$$

by

$$\begin{cases} \mathrm{Glures}_{1,S}(T,\rho)(x) & = u_T^\rho(x) \qquad x \in K_1 \\ \mathrm{Glures}_{1,S}(T,\rho)(\tau',t) & = u_T^\rho(\tau',t) \;\; (= u_T^\rho(\tau + 5T, t)) \end{cases} \tag{6.5}$$

$$\begin{cases} \mathrm{Glures}_{2,S}(T,\rho)(x) & = u_T^\rho(x) \;\; x \in K_2 \\ \mathrm{Glures}_{2,S}(T,\rho)(\tau'',t) & = u_T^\rho(\tau'',t) \quad (= u_T^\rho(\tau - 5T, t)). \end{cases} \tag{6.6}$$

Here the map 'Glures' stands for the phrase 'gluing followed by restriction', and $\mathrm{Map}_{L^2_{m+1}}((K_i^S, K_i^S \cap \partial\Sigma_i), (X, L))$ is the space of maps of L^2_{m+1} class (m is sufficiently large, say $m > 10$.) It has the structure of Hilbert manifold in an obvious way. This Hilbert manifold is independent of T. So we can define the T derivative of a family of elements of $\mathrm{Map}_{L^2_{m+1}}((K_i^S, K_i^S \cap \partial\Sigma_i), (X, L))$ parameterized by T.[1]

[1] We can also use $C^k((K_i^S, K_i^S \cap \partial\Sigma_i), (X, L))$, the Banach manifold of C^k maps, in place of $\mathrm{Map}_{L^2_{m+1}}((K_i^S, K_i^S \cap \partial\Sigma_i), (X, L))$.

REMARK 6.3. The domain and the codomain of the map $\mathrm{Glures}_{i,S}$ depend on m. However its image actually lies in the set of smooth maps. Also none of the constructions of u^{ρ}_T depend on m. (The proof of the convergence of (6.1) depends on m. So the numbers $T_{3,m,\epsilon(5)}$, $T_{4,m}$ in Proposition 5.23 (and $T_{5,m,\epsilon(6)}$, $T_{6,m}$ in Proposition 6.9) depend on m.) Therefore the map $\mathrm{Glures}_{i,S}$ is *independent* of m on the intersection of the domains. Namely the map $\mathrm{Glures}_{i,S}$ constructed by using the $L^2_{m_1}$ norm coincides with the map $\mathrm{Glures}_{i,S}$ constructed by using the $L^2_{m_2}$ norm on $[T', \infty) \times V_1 \times_L V_2$ where

$$T' = \max\{T_{3,m_1,\epsilon(5)}, T_{4,m_1}, T_{5,m_1,\epsilon(6)}, T_{6,m_1} T_{3,m_2,\epsilon(5)}, T_{4,m_2}, T_{5,m_2,\epsilon(6)}, T_{6,m_2}\}.$$

THEOREM 6.4. *For each given m and S, there exist $T_{m,S,(6.7)} > S$, $C_{m,S,(6.7)} > 0$ such that*

$$\left\| \nabla^n_{\rho} \frac{d^{\ell}}{dT^{\ell}} \mathrm{Glures}_{i,S} \right\|_{L^2_{m+1-\ell}} < C_{m,S,(6.7)} e^{-\delta_1 T} \tag{6.7}$$

holds for all $T > T_{m,S,(6.7)}$ and for (n, ℓ) with $m-2 \geq n, \ell \geq 0$ and $\ell > 0$. Here ∇^n_{ρ} is the n-th derivative in the ρ-direction.

REMARK 6.5. Theorem 6.4 is equivalent to [**FOOO1**, Lemma A1.58]. The proof below uses the same inductive scheme as the one in [**FOOO1**, page 776]. We use the L^2_m norm in place of the L^p_1 norm and add thorough detail.

REMARK 6.6. We remark that the map

$$\mathcal{M}^{\mathcal{E}_1 \oplus \mathcal{E}_2}((\Sigma_T, \vec{z}); u_1, u_2)_{\epsilon_1} \to \prod_{i=1}^{2} \mathrm{Map}_{L^2_{m+1}}((K^S_i, K^S_i \cap \partial \Sigma_i), (X, L)),$$

which is obtained by restricting the domain, is an embedding for each fixed T. This is a consequence of unique continuation. We note that the codomain of the map does not depend on T. Therefore we can use Theorem 6.4 to study T-dependence of the moduli space $\mathcal{M}^{\mathcal{E}_1 \oplus \mathcal{E}_2}((\Sigma_T, \vec{z}); \beta)_{\epsilon}$. We used this fact and Theorem 6.4 to show smoothness of the coordinate change of the Kuranishi structure on the moduli space of bordered pseudoholomorphic curves in [**FOOO1**, Appendix A, page 764–773]. See Chapter 8 for more details.

The remaining chapter will be occupied by the proof of this theorem.

The construction of $u^{\rho}_{T,(\kappa)}$ was given by induction on κ. We divide the inductive step of the construction of $u^{\rho}_{T,(\kappa)}$ from $u^{\rho}_{T,(\kappa-1)}$ into two:

(Part A) Start from

$$(V^{\rho}_{T,1,(\kappa)}, V^{\rho}_{T,2,(\kappa)}, \Delta p^{\rho}_{T,(\kappa)})$$

and end with $\mathrm{Err}^{\rho}_{1,T,(\kappa)}$ and $\mathrm{Err}^{\rho}_{2,T,(\kappa)}$. This is the steps κ-2,κ-3 and κ-4. This is the step of the error estimates for the κ-th iteration map $u^{\rho}_{T,(\kappa)}$, which is essentially a computational step.

(Part B) Start from $\mathrm{Err}^{\rho}_{1,T,(\kappa-1)}$ and $\mathrm{Err}^{\rho}_{2,T,(\kappa-1)}$ and end with

$$(V^{\rho}_{T,1,(\kappa)}, V^{\rho}_{T,2,(\kappa)}, \Delta p^{\rho}_{T,(\kappa)}).$$

This is the step κ-1. This step involves inverting the approximate right inverse of the linearization of the equation (3.19) at $u^{\rho}_{T,(\kappa-1)}$.

We denote by

$$\mathscr{P}_{i,\rho_i} : (u_i^{\rho_i})^*TX \to u_i^*TX, \quad (\text{resp. } \mathcal{P}^{p_0}_{p_0^\rho} : T_{p_0^\rho}X \to T_{p_0}X) \tag{6.8}$$

the bundle map (resp. the map) induced by the parallel transportation along the minimal geodesic. The main task of this chapter is to prove the following inequalities inductively over $\kappa \geq 0$:

$$\left\| \nabla_\rho^n \frac{\partial^\ell}{\partial T^\ell}(V^\rho_{T,i,(\kappa)}, \Delta p^\rho_{T,(\kappa)}) \right\|_{W^2_{m+1-\ell,\delta}(\Sigma_i)} < C_{5,m}\mu^{\kappa-1}e^{-\delta_1 T}, \tag{6.9}$$

$$\begin{aligned} &\left\| \nabla_\rho^n \frac{\partial^\ell}{\partial T^\ell}(\mathscr{P}_{i,\rho_i}(\mathrm{E}(u_i^{\rho_i}, u^\rho_{T,(\kappa)})), \mathcal{P}^{p_0}_{p_0^\rho}(\mathrm{E}(p_0^\rho, p^\rho_{T,(\kappa)}))) \right\|_{W^2_{m+1-\ell,\delta}(K_i^{5T+1}\subset\Sigma_i)} \\ &\qquad < C_{6,m}(2-\mu^\kappa)e^{-\delta_1 T}, \end{aligned} \tag{6.10}$$

$$\left\| \nabla_\rho^n \frac{\partial^\ell}{\partial T^\ell}\mathrm{E}(p_0^\rho, u^\rho_{T,(\kappa)}) \right\|_{L^2_{m+1-\ell}(K_i^{9T}\setminus K_i^T)} < C_{7,m}(2-\mu^\kappa)e^{-\delta_1 T}, \tag{6.11}$$

$$\left\| \nabla_\rho^n \frac{\partial^\ell}{\partial T^\ell}\mathrm{Err}^\rho_{i,T,(\kappa)} \right\|_{L^2_{m-\ell,\delta}(\Sigma_i)} < C_{8,m}\epsilon(6)\mu^\kappa e^{-\delta_1 T}, \tag{6.12}$$

$$\left\| \nabla_\rho^n \frac{\partial^\ell}{\partial T^\ell}\mathfrak{e}^\rho_{i,T,(\kappa)} \right\|_{L^2_{m-\ell}(K_i^{\mathfrak{ob}})} < C_{9,m}\mu^{\kappa-1}e^{-\delta_1 T}. \tag{6.13}$$

Here $0 \leq \ell, n \leq m-2$.

REMARK 6.7.

(1) Note we use the T-independent coordinates (τ', t) on $K_1^{5T+1}\setminus K_1$, $K_1^{9T}\setminus K_1^T$ and (τ'', t) on $K_2^{5T+1}\setminus K_2$, $K_2^{9T}\setminus K_2^T$.
(2) In (6.11), (6.12) we use the τ' coordinate for $i=1$ and the τ'' coordinate for $i=2$.
(3) We also remark that $\Sigma_T = K_1^{5T+1} \cup K_2^{5T+1}$. (See the definition of Σ_T at the beginning of this chapter and (6.4).)
(4) Note (6.11) implies the same inequality with $u^\rho_{T,(\kappa)}$ replaced by $\hat u^\rho_{i,T,(\kappa)}$, which is defined in (5.74).

Some explanation of the derivatives appearing in (6.9)-(6.13) is also needed. Recall from Definition 5.31 that the pair $(V^\rho_{T,i,(\kappa)}, \Delta p^\rho_{T,(\kappa)})$ appearing in (6.9) is an element of the weighted Sobolev space

$$W^2_{m+1,\delta}((\Sigma_i,\partial\Sigma_i);(\hat u^\rho_{i,T,(\kappa-1)})^*TX,(\hat u^\rho_{i,T,(\kappa-1)})^*TL)$$

which depends on T and ρ. We use the inverse of parallel transport map

$$\Phi_{i,(\kappa-1)}(\rho,T)$$

(Definition 5.28), to make the identification

$$\begin{aligned} &W^2_{m+1,\delta}((\Sigma_i,\partial\Sigma_i);(\hat u^\rho_{i,T,(\kappa-1)})^*TX,(\hat u^\rho_{i,T,(\kappa-1)})^*TL) \\ &\cong W^2_{m+1,\delta}((\Sigma_i,\partial\Sigma_i);u_i^*TX,u_i^*TL). \end{aligned}$$

where the latter space is (T,ρ)-independent. Namely we put

$$\Psi_{i;(\kappa-1)}(\rho,T;(s,v)) = \left((\Phi_{i;(\kappa-1)}(\rho,T)^{-1}(s), (\mathrm{Pal}_{p_0}^{p^\rho_{(\kappa-1)}})^{-1}(v) \right). \tag{6.14}$$

Through this identification, the precise meaning of the formula (6.9) is the inequality:

$$\left\| \nabla_\rho^n \frac{\partial^\ell}{\partial T^\ell}(\Psi_{i;(\kappa-1)}(\rho, T; (V^\rho_{T,i,(\kappa)}, \Delta p^\rho_{T,(\kappa)}))) \right\|_{W^2_{m+1-\ell,\delta}(\Sigma_i)} \le C_{5,m}\mu^{\kappa-1}e^{-\delta_1 T}. \tag{6.15}$$

We use the τ' coordinate in case $i = 1$.

We can make sense of (6.12) in the same way as (6.15). We use the isomorphism

$$L^2_{m,\delta}(\Sigma_i; (\hat{u}^\rho_{i,T,(\kappa-1)})^* TX \otimes \Lambda^{0,1}) \cong L^2_{m,\delta}(\Sigma_i; u_i^* TX \otimes \Lambda^{0,1}), \tag{6.16}$$

to make sense out of (6.13). The isomorphism is nothing but the closure of

$$\left((\mathrm{Pal}_{u_i}^{\hat{u}^\rho_{i,T,(\kappa-1)}})^{(0,1)}\right)^{-1},$$

where $(\mathrm{Pal}_{u_i}^{\hat{u}^\rho_{i,T,(\kappa-1)}})^{(0,1)}$ is as in (2.7), to formulate (6.13).

A similar remark applies to (6.11). (6.11) means

$$\left\| \nabla_\rho^n \frac{\partial^\ell}{\partial T^\ell} \mathrm{Pal}_{p_0^\rho}^{p_0}(\mathrm{E}(p_0^\rho, u^\rho_{T,(\kappa)})) \right\|_{L^2_{m+1-\ell}(K_i^{9T}\setminus K_i^T)} \le C_{7,m}(2-\mu^\kappa)e^{-\delta_1 T} \tag{6.17}$$

and in (6.13) we regard $\mathfrak{e}^\rho_{i,T,(\kappa)} \in \mathcal{E}_i^{\mathfrak{ob}}$.

In this way we can safely work with (6.9)-(6.13).

REMARK 6.8. Similar remarks also apply to the case $i = 2$ using the τ'' coordinate.

The inductive proof of (6.9)-(6.13) is written as the proof of the next proposition.

PROPOSITION 6.9. *We can choose $C_{5,m}$, $C_{8,m}$, $C_{9,m}$ independent of $\kappa \ge 1$ so that the following holds:*

(1) *For any $\epsilon(6) > 0$ and $C_{6,m}$, $C_{7,m}$, there exists $T_{5,m,\epsilon(6)} > 0$ such that (6.9)-(6.11) for $\kappa \le \kappa_0$ imply (6.12) and (6.13) for $\kappa \le \kappa_0$ for all $T > T_{5,m,\epsilon(6)}$.*
(2) *For any given $C_{6,m}$, $C_{7,m}$, we can choose $\epsilon_{4,m}$ so that if $\epsilon(6) < \epsilon_{4,m}$ then (6.9)-(6.13) for $\kappa \le \kappa_0 - 1$ imply (6.9) for κ_0.*
(3) *We can choose $C_{6,m}$, $C_{7,m}$ with the following properties: There exists $T_{6,m}$ such that the inequalities (6.10), (6.11) for κ_0 follow from (6.9) for $\kappa \le \kappa_0$ and (6.10), (6.11) for $\kappa \le \kappa_0 - 1$ if $T > T_{6,m}$.*

REMARK 6.10. The constants $T_{5,m,\epsilon(6)}$, $\epsilon_{4,m}$, $T_{6,m}$ may depend on $C_{5,m}$, $C_{6,m}$, $C_{7,m}$, $C_{8,m}$, $C_{9,m}$.

The rest of this chapter will be occupied by the proof of Proposition 6.9. (3) is elementary. We provide its proof in Appendix E for completeness' sake. We will prove (1) and (2) in this chapter. We divide our proof into two parts, the proofs of (1) and of (2).

REMARK 6.11. We choose the constants $C_{5,m}$, $C_{8,m}$, $C_{9,m}$ so that (6.9), (6.12) and (6.13) hold for $\kappa = 0, 1$. We do not need to change them in the later steps. The constants $C_{6,m}$, $C_{7,m}$ are chosen during the proof of Proposition 6.9 (3) given in Appendix E.

LEMMA 6.12. *Let $\kappa_0 \geq 1$ be given. The inequalities* (6.9), (6.10), (6.11), (6.12) *and* (6.13) *for $\kappa \leq \kappa_0$ imply*

$$\left\| \nabla^n_\rho \frac{\partial^\ell}{\partial T^\ell} \mathrm{E}(u_i, u^\rho_{T,(\kappa_0)}) \right\|_{W^2_{m+1-\ell,\delta}(K_i^{5T+1} \subset \Sigma_i)} \begin{cases} \leq C_{m,(6.18)} & \ell = 0 \\ \leq C'_{m,(6.18)} e^{-T\delta_1} & \ell > 0 \end{cases} \tag{6.18}$$

for $0 \leq n \leq m-2$, $0 \leq \ell \leq m-2$.

PROOF. The case $\ell > 0$ follows from (6.10) by using the fact $u_i^{\rho_i}$ is independent of T. Using the fact that the C^∞ norm between u_i and $u_i^{\rho_i}$ is uniformly bounded and small as ρ_i moves, we can use (6.10) to prove the case $\ell = 0$ of (6.18) by using Lemma E.3. □

6.2. Part A: Error estimates

In this section we prove Proposition 6.9 (1). This section corresponds to the discussion in [**FOOO1**, page 776 paragraph (A) and (B)].

Suppose that the triple $(V^\rho_{T,1,(\kappa)}, V^\rho_{T,2,(\kappa)}, \Delta p^\rho_{T,(\kappa)})$ satisfies (6.9). We denote

$$\mathrm{Err}^\rho_{T,(\kappa)} = \overline{\partial} u^\rho_{T,(\kappa)} - \sum_{a=0}^{\kappa} \mathfrak{e}^\rho_{1,T,(a)} - \sum_{a=0}^{\kappa} \mathfrak{e}^\rho_{2,T,(a)} \tag{6.19}$$

and

$$\mathrm{Err}^\rho_{1,T,(\kappa)} = \chi^{\leftarrow}_{\mathcal{X}} \mathrm{Err}^\rho_{T,(\kappa)}, \qquad \mathrm{Err}^\rho_{2,T,(\kappa)} = \chi^{\rightarrow}_{\mathcal{X}} \mathrm{Err}^\rho_{T,(\kappa)}.$$

Then noting that $\operatorname{supp} \mathfrak{e}_{i,T,(\kappa)} \subset \operatorname{Int} K_i$, we find that

(1)

$$\mathrm{Err}^\rho_{1,T,(\kappa)}(z) = \Pi^\perp_{\mathcal{E}_1(\hat{u}^\rho_{1,T,(\kappa-1)})} \overline{\partial} \left(\mathrm{Exp}\left(\hat{u}^\rho_{1,T,(\kappa-1)}(z), V^\rho_{T,1,(\kappa)}(z) \right) \right) \tag{6.20}$$

for $z \in K_1$.

(2)

$$\begin{aligned} &\mathrm{Err}^\rho_{1,T,(\kappa)}(\tau', t) \\ &= (1 - \chi(\tau' - 5T)) \times \\ &\overline{\partial}\Big(\mathrm{Exp}\Big(u^\rho_{T,(\kappa-1)}(\tau', t), \chi(\tau' - 4T) \Big(V^\rho_{T,2,(\kappa)}(\tau' - 10T, t) - (\Delta p^\rho_{T,(\kappa)})^{\mathrm{Pal}} \Big) \\ &\qquad + V^\rho_{T,1,(\kappa)}(\tau', t) \Big) \Big) \end{aligned} \tag{6.21}$$

for $(\tau', t) \in [0, \infty)_{\tau'} \times [0, 1]$. (Recall $\tau = \tau' - 5T$, $\tau' = \tau'' - 10T$ and $V^\rho_{T,2,(\kappa)}$ was defined in terms of the variable of (τ'', t).) See Figure 6.2.

Here $\chi : \mathbb{R} \to [0, 1]$ is a smooth function such that

$$\chi(\tau) = \begin{cases} 0 & \tau < -1 \\ 1 & \tau > 1 \end{cases}$$

$$\chi'(\tau) > 0 \text{ for } \tau \in (-1, 1). \tag{6.22}$$

The same kind of statements also hold for $i = 2$. Since the latter case can be dealt exactly in the same way, we omit its details.

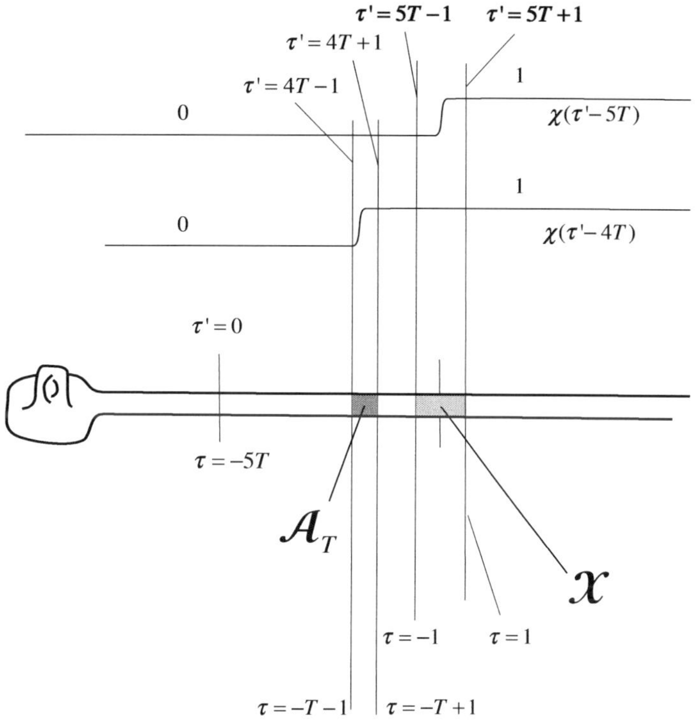

FIGURE 6.2. Cut off function.

REMARK 6.13. Note that in Formulae (6.9)-(6.13) the Sobolev norms in the left hand side are of $W^2_{m+1-\ell,\delta}(\Sigma_i)$ etc. and are not of $W^2_{m+1,\delta}(\Sigma_i)$ etc. The origin of this loss of differentiability (in the sense of Sobolev space) comes from the term $V^\rho_{T,2,(\kappa)}(\tau' - 10T)$. In fact, we have

$$\frac{\partial}{\partial T} V^\rho_{T_1,2,(\kappa)}(\tau' - 10T) = -10 \frac{\partial V^\rho_{T_1,2,(\kappa)}}{\partial \tau''}(\tau' - 10T)$$

for a fixed T_1. Hence $\partial/\partial T$ is continuous as $L^2_{m+1} \to L^2_m$. We remark in (6.9) for $i = 2$ we use the coordinates (τ'', t) on $(-\infty, 0] \times [0,1]$ to define the T derivative of $V^\rho_{T,2,(\kappa)}$.

Taking this remark into account, we continue with the proof. The proof is divided into three parts.

(1) On $K_1^{4T-1} \cup K_2^{4T-1}$, where we denote[2]

$$K_1^{4T-1} = K_1 \cup [0, 4T-1]_{\tau'} \times [0,1],$$
$$K_2^{4T-1} = K_2 \cup [-4T+1, 0]_{\tau''} \times [0,1]. \tag{6.23}$$

(2) On the central neck region containing $\mathcal{X} = [-1,1]_\tau \times [0,1]$.

(3) On the transition region

$$([4T-1, 4T+1]_{\tau'} \times [0,1)) \cup ([-4T-1, -4T+1]_{\tau''} \times [0,1]).$$

[2]The notation $[0, 4T-1]_{\tau'}$ is introduced in Remark 6.1.

(Estimate 1): Let K_1^{4T-1}, K_2^{4T-1} be as in (6.23). We remark that on $K_1^{4T-1} \setminus K_1$ the formula (6.21) is reduced to

$$\mathrm{Err}^{\rho}_{1,T,(\kappa)}(\tau',t) = \overline{\partial}\Big(\mathrm{Exp}\left(u^{\rho}_{T,(\kappa-1)}(\tau',t), V^{\rho}_{T,1,(\kappa)}(\tau',t)\right)\Big) \tag{6.24}$$

since $1-\chi(\tau'-5T) \equiv 1$ as $\tau'-5T \le -T+1 < -1$ and $\chi(\tau'-4T) \equiv 0$ as $\tau'-4T \le -1$. Using the fact that $V^{\rho}_{T,2,(\kappa)}$ does not appear in (6.24), we can estimate T and ρ derivatives of $\mathrm{Err}^{\rho}_{1,T,(\kappa)}$ on K_1^{4T-1}, K_2^{4T-1} by the same way as the corresponding part (that is, Formula (C.10)) of Proposition 5.34 given in Appendix C as follows.

LEMMA 6.14. *For any $\epsilon(7) > 0$, there exists $T_{m,\epsilon(7),(6.25)} > 0$ with the following properties. If $T > T_{m,\epsilon(7),(6.25)}$, then the element $\mathfrak{e}^{\rho}_{i,T,(\kappa)} \in \mathcal{E}_i$ in Proposition 5.34 satisfies the following for $0 \le \ell, n \le m-2$.*

$$\left\| \nabla^n_{\rho} \frac{d^{\ell}}{dT^{\ell}} \left((\mathrm{Pal}_{u_i}^{u^{\rho}_{T,(\kappa)}})^{(0,1)} \right)^{-1} \left(\mathrm{Err}^{\rho}_{T,(\kappa)} \right) \right\|_{L^2_{m-\ell,\delta}(K_i^{4T-1})} \le C_{5,m}\mu^{\kappa}\epsilon(7)e^{-\delta_1 T}, \tag{6.25}$$

and

$$\left\| \nabla^n_{\rho} \frac{d^{\ell}}{dT^{\ell}} \left(\mathfrak{e}^{\rho}_{i,T,(\kappa)} \right) \right\|_{L^2_{m-\ell}(K_i^{4T-1})} < \frac{C_{8,m}}{10}\mu^{\kappa-1}e^{-\delta_1 T}. \tag{6.26}$$

We provide its proof in Appendix G for completeness.

We next study the neck region. The point explained in Remark 6.13 appears here. We will do the corresponding estimates for $\mathrm{Err}^{\rho}_{1,T,(\kappa)}$ given in (6.21) by considering them for $\tau' \in [4T+1,\infty)_{\tau'}$ and for $\tau' \in [4T-1, 4T+1]_{\tau'}$, separately.
(Estimate 2): We first consider the domain $(\tau',t) \in [4T+1,\infty)_{\tau'} \times [0,1]_t$. (Note this domain contains $\mathfrak{X}$.) We put:

$$\begin{aligned} h^0(\tau',t) &= u^{\rho}_{T,(\kappa-1)}(\tau',t) \\ h^1(\tau',t) &= \mathrm{Exp}\Big(u^{\rho}_{T,(\kappa-1)}(\tau',t), \\ &\qquad V^{\rho}_{T,2,(\kappa)}(\tau'-10T,t) - (\Delta p^{\rho}_{T,(\kappa)})^{\mathrm{Pal}} + V^{\rho}_{T,1,(\kappa)}(\tau',t)\Big). \end{aligned} \tag{6.27}$$

Then for $\tau' \in [4T+1,\infty)_{\tau'}$ the formula (6.21) is reduced to

$$\mathrm{Err}^{\rho}_{1,T,(\kappa)}(\tau',t) = (1-\chi(\tau'-5T))\overline{\partial}h^1, \tag{6.28}$$

because $\chi(\tau'-4T) = 1$ on $[4T+1,\infty)$ as $\tau'-4T \ge 1$.

We remark that $\mathrm{Err}^{\rho}_{1,T,(\kappa)}(\tau',t) = 0$ on $\tau' > 5T+1$ since $1-\chi(\tau'-5T) = 0$ there. So we need to study only on $\tau' \in [4T+1, 5T+1]_{\tau'}$. To estimate this error term, we consider a family of maps h^r such that $r \mapsto h^r(\tau',t)$ is the minimal geodesic joining $h^0(\tau',t)$ and $h^1(\tau',t)$. Namely

$$h^r(\tau',t) = \mathrm{Exp}(h^0(\tau',t), r\mathrm{E}(h^0(\tau',t), h^1(\tau',t))).$$

More explicitly:

$$\begin{aligned} h^r(\tau',t) &= \mathrm{Exp}\Big(u^{\rho}_{T,(\kappa-1)}(\tau',t), \\ &\qquad r(V^{\rho}_{T,2,(\kappa)}(\tau'-10T,t) - (\Delta p^{\rho}_{T,(\kappa)})^{\mathrm{Pal}} + V^{\rho}_{T,1,(\kappa)}(\tau',t))\Big). \end{aligned} \tag{6.29}$$

We again apply (5.42) pointwise to the function

$$g(s) = \mathcal{P}^{-1}(\overline{\partial} h^s)$$

where $\mathcal{P}$ is induced from the parallel transport along the curve $r \mapsto h^r(\tau', t)$, $r \in [0, s]$ at each given point (τ', t). We also note

$$g(1)(\tau', t) = \frac{1}{(1 - \chi(\tau' - 5T))} \mathcal{P}^{-1}(\mathrm{Err}^{\rho}_{1,T,(\kappa)}(\tau', t)).$$

We remark that

$$\begin{aligned} g(0)(\tau', t) &= \overline{\partial} u^{\rho}_{T,(\kappa-1)}(\tau', t), \\ g'(0)(\tau', t) &= \left((D_{h^0}\overline{\partial})(\mathrm{E}(h^0, h^1))\right)(\tau', t). \end{aligned} \tag{6.30}$$

We denote the parallel transport along the shortest path by

$$\mathfrak{P} = \left(\mathrm{Pal}^{h^0}_{u_1}\right)^{(0,1)}$$

on the domain $[0, 5T]_{\tau'} \times [0, 1]$.

We first examine $g(0)$. By (5.72) we have

$$\overline{\partial} h^0(\tau', t) = \overline{\partial} u^{\rho}_{T,(\kappa-1)}(\tau', t) = \mathrm{Err}^{\rho}_{1,T,(\kappa-1)}(\tau', t) + \mathrm{Err}^{\rho}_{2,T,(\kappa-1)}(\tau', t)$$

outside the supports of $\mathcal{E}_1$, $\mathcal{E}_2$, which contain the subset $[4T + 1, 5T + 1]_{\tau'} \times [0, 1]$ we are studying.

Next we examine the term $g'(0)$. The second formula of (6.30) implies:

$$\begin{aligned} g'(0)(\tau', t) = (D_{u^{\rho}_{T,(\kappa-1)}} \overline{\partial})\big(&V^{\rho}_{T,2,(\kappa)}(\tau' - 10T, t) \\ &+ V^{\rho}_{T,1,(\kappa)}(\tau', t) - (\Delta p^{\rho}_{T,(\kappa)})^{\mathrm{Pal}}\big). \end{aligned} \tag{6.31}$$

Using (6.18) and (6.12) we obtain

$$\begin{aligned} &\left\| \nabla^n_{\rho} \frac{d^{\ell}}{dT^{\ell}} \mathfrak{P}^{-1} (D_{u^{\rho}_{T,(\kappa-1)}} \overline{\partial})\big((\Delta p^{\rho}_{T,(\kappa)})^{\mathrm{Pal}}\big) \right\|_{L^2_{m-\ell}([T,9T]_{\tau'} \times [0,1])} \\ &\le \sum_{\ell' \le \ell, n' \le n} C_{m,(6.32)} T e^{-\delta_1 T} \left\| \nabla^{n'}_{\rho} \frac{d^{\ell'}}{dT^{\ell'}} \Delta p^{\rho}_{T,(\kappa)} \right\|_{L^2_{m-\ell'}([T,9T]_{\tau'} \times [0,1])} \\ &\le C'_{m,(6.32)} T \mu^{\kappa-1} e^{-2\delta_1 T}. \end{aligned} \tag{6.32}$$

To show this inequality for $\ell' = \ell$ we use the fact that $(\Delta p^{\rho}_{T,(\kappa)})^{\mathrm{Pal}}$ is almost a 'constant' and its first derivative is small. See the last part of Appendix G.

Moreover by (5.85) in Definition 5.31 and Definition 5.35, we have

$$-(D_{\hat{u}^{\rho}_{i,T,(\kappa-1)}} \overline{\partial})\big(V^{\rho}_{T,i,(\kappa)}\big) = \mathrm{Err}^{\rho}_{i,T,(\kappa-1)}(\tau', t) \tag{6.33}$$

for $i = 1, 2$, on $[4T + 1, 5T + 1]_{\tau'} \times [0, 1]$, which lies outside the supports of $\mathcal{E}_1$, $\mathcal{E}_2$. Note $\hat{u}^{\rho}_{i,T,(\kappa-1)} = u^{\rho}_{T,(\kappa-1)}$ on the domain $[4T + 1, 5T + 1]_{\tau'} \times [0, 1]$, since $\chi_{\mathcal{A}}$ and $\chi_{\mathcal{B}}$ are 1 therein. (See (5.74).)

Finally we establish the following estimate.

LEMMA 6.15. *Let g be as above. For any $\epsilon(8) > 0$ there exists $T_{m,\epsilon(8),(6.34)}$ such that*

$$\left\| \nabla^n_{\rho} \frac{d^{\ell}}{dT^{\ell}} \left(\int_0^1 \left(\int_0^s g''(r) dr \right) ds \right) \right\|_{L^2_{m-\ell,\delta}([4T+1,5T+1] \times [0,1])} \le \mu^{\kappa} \epsilon(8) e^{-\delta_1 T} \tag{6.34}$$

holds for all $T > T_{m,\epsilon(8),(6.34)}$.

PROOF. We put

$$V(\tau',t) = (\mathfrak{P}')^{-1}\left((V^{\rho}_{T,2,(\kappa)}(\tau'-10T,t) - (\Delta p^{\rho}_{T,(\kappa)})^{\mathrm{Pal}}) + V^{\rho}_{T,1,(\kappa)}(\tau',t)\right),$$

where

$$\mathfrak{P}' = \mathrm{Pal}^{h^0}_{u_1}.$$

(Note on the domain $[4T+1,5T+1]_{\tau'} \times [0,1]$, we have $\hat{u}^{\rho}_{2,T,(\kappa-1)} = u^{\rho}_{T,(\kappa-1)}$. So we may regard $V^{\rho}_{T,2,(\kappa)}$ as a section of $(u^{\rho}_{T,(\kappa-1)})^*TX$.)

We take the $L^2_{m-\ell}$ norm in place of the $L^2_{m-\ell,\delta}$ norm of the left hand side of (6.34). Then in the same way as in the proof of Lemma 6.14 given in Appendix G we can estimate the norm by

$$C_{m,(6.35)} \sum_{\substack{\ell_1+\ell_2\le \ell\\ n_1+n_2\le n}} \left\Vert \nabla_{\rho}^{n_1}\frac{d^{\ell_1}}{dT^{\ell_1}}V\right\Vert_{L^2_{m+1-\ell_1}([4T+1,5T+1]\times[0,1])} \times \left\Vert \nabla_{\rho}^{n_2}\frac{d^{\ell_2}}{dT^{\ell_2}}V\right\Vert_{L^2_{m+1-\ell_2}([4T+1,5T+1]\times[0,1])}. \tag{6.35}$$

Then applying the formula

$$\begin{aligned}&\frac{\partial^{\ell}}{\partial T^{\ell}}(\mathfrak{P}')^{-1}\left(V^{\rho}_{T,2,(\kappa)}(\tau'-10T)\right)\Big|_{T=T_2}\\ &= \sum_{\ell_1+\ell_2=\ell}(-10)^{\ell_2}\frac{\partial^{\ell_1}}{\partial T^{\ell_1}}\frac{\partial^{\ell_2}((\mathfrak{P}')^{-1}(V^{\rho}_{T,2,(\kappa)}))}{\partial \tau''^{\ell_2}}(\tau'-10T_2),\end{aligned} \tag{6.36}$$

we obtain:

$$\begin{aligned}&\left\Vert \nabla_{\rho}^{n_i}\frac{d^{\ell_i}}{dT^{\ell_i}}V\right\Vert_{L^2_{m+1-\ell_i}([4T+1,5T+1]_{\tau'}\times[0,1])}\\ &\le C_{m,(6.37)}\left\Vert \nabla_{\rho}^{n_i}\frac{d^{\ell_i}}{dT^{\ell_i}}(\mathfrak{P}')^{-1}(V^{\rho}_{T,1,(\kappa)})\right\Vert_{L^2_{m+1-\ell_i}([4T+1,5T+1]_{\tau'}\times[0,1])}\\ &\quad + C_{m,(6.37)}\left\Vert \nabla_{\rho}^{n_i}\frac{d^{\ell_i}}{dT^{\ell_i}}(\mathfrak{P}')^{-1}(V^{\rho}_{T,2,(\kappa)})\right\Vert_{L^2_{m+1}([-6T_1+1,-5T+1]_{\tau''}\times[0,1])}\\ &\quad + C_{m,(6.37)}\left\Vert \nabla_{\rho}^{n_i}\frac{d^{\ell_i}}{dT^{\ell_i}}(\mathfrak{P}')^{-1}((\Delta p^{\rho}_{T,(\kappa)})^{\mathrm{Pal}})\right\Vert_{L^2_{m+1-\ell_i}([4T+1,5T+1]_{\tau'}\times[0,1])}\\ &\le C'_{m,(6.37)}\mu^{\kappa-1}e^{-\delta_1 T}.\end{aligned} \tag{6.37}$$

Note the weight function on our domain $[4T+1,5T+1]\times[0,1]$ is not greater than $10e^{5T\delta}$. Therefore substituting (6.37) into (6.35), we obtain

$$\text{LHS of (6.34)} \le C_{m,(6.38)}e^{5T\delta}e^{-2T\delta_1}\mu^{2\kappa-2} \le \mu^{\kappa}\epsilon(8)e^{-\delta_1 T} \tag{6.38}$$

by taking $T_{m,\epsilon(8),(6.34)}$ so that $C_{m,(6.38)}e^{-5T_{m,\epsilon(8),(6.34)}\delta}\mu^{\kappa-2} \le \epsilon(8)$. Here we also used (3.11). □

We use Lemma 6.15 and (6.31), (6.32), (6.33) to show

$$
\begin{aligned}
&\left\|\nabla_\rho^n \frac{d^\ell}{dT^\ell}\mathfrak{P}^{-1}\mathrm{Err}^\rho_{1,T,(\kappa)}\right\|_{L^2_{m-\ell,\delta}([4T+1,5T+1]_{\tau'}\times[0,1])} \\
&=\left\|\nabla_\rho^n \frac{d^\ell}{dT^\ell} g(1)\right\|_{L^2_{m-\ell,\delta}([4T+1,5T+1]_{\tau'}\times[0,1])} \\
&\le \left\|\nabla_\rho^n \frac{d^\ell}{dT^\ell}(g(0)+g'(0))\right\|_{L^2_{m-\ell,\delta}([4T+1,5T+1]_{\tau'}\times[0,1])} + \text{LHS of (6.34)} \\
&\le C'_{m,(6.32)} T\mu^{\kappa-1}e^{-2\delta_1 T} + \mu^\kappa \epsilon(8)e^{-\delta_1 T} \\
&\le 2\mu^\kappa\epsilon(8)e^{-\delta_1 T},
\end{aligned}
\tag{6.39}
$$

for $T > T_{m,\epsilon(8),(6.34)}$, by taking $C'_{m,(6.32)}T\mu^{\kappa-1}e^{-\delta_1 T} \le \mu^\kappa\epsilon(8)$.

(Estimate 3): We next consider $\tau' \in [4T-1, 4T+1]_{\tau'}$. In other words we study the estimate on the domain $\mathcal{A}_T$. There the formula (6.21) for the error term is reduced to

$$
\begin{aligned}
&\mathrm{Err}^\rho_{1,T,(\kappa)}(\tau',t) \\
&=\overline{\partial}\Big(\mathrm{Exp}\left(u^\rho_{T,(\kappa-1)}(\tau',t), V^\rho_{T,1,(\kappa)}(\tau',t)+\right. \\
&\qquad\qquad \left.\chi(\tau'-4T)\Big(V^\rho_{T,2,(\kappa)}(\tau'-10T,t) - (\Delta p^\rho_{T,(\kappa)})^{\mathrm{Pal}}\Big)\right)\Big)
\end{aligned}
\tag{6.40}
$$

since $1-\chi(\tau'-5T)\equiv 1$ as $\tau'-5T \le -T+1 < -1$.

We consider the path of maps $h^s(\tau',t)$ given by

$$
\begin{aligned}
h^s(\tau',t) = \mathrm{Exp}&\left(u^\rho_{T,(\kappa-1)}(\tau',t), V^\rho_{T,1,(\kappa)}(\tau',t)+\right. \\
&\qquad \left. s\chi(\tau'-4T)\Big(V^\rho_{T,2,(\kappa)}(\tau'-10T,t) - (\Delta p^\rho_{T,(\kappa)})^{\mathrm{Pal}}\Big)\right)
\end{aligned}
\tag{6.41}
$$

which satisfies

$$
\begin{aligned}
\overline{\partial}h^0 &= \overline{\partial}\,\mathrm{Exp}\left(u^\rho_{T,(\kappa-1)}(\tau',t), V^\rho_{T,1,(\kappa)}(\tau',t)\right) \\
\overline{\partial}h^1 &= \mathrm{Err}^\rho_{1,T,(\kappa)}.
\end{aligned}
$$

The next lemma claims that the term containing $V^\rho_{T,2,(\kappa)}(\tau'-10T,t)$ in (6.40) is small.

Let $(\mathcal{P}')^{-1}$ denote the parallel transport along the path $s\mapsto h^s(\tau',t)$ ($s\in[0,r]$) and $\mathcal{P}^{-1}$ its complex linear part. We denote by $\mathfrak{P}'$ the parallel transport along the minimal geodesic which maps a section of u_1^*TX to a section of $(u')^*TX$ for various u' which is C^0 close to u_1. Let $\mathfrak{P}$ be its complex linear part.

LEMMA 6.16. *Let h^s be as above. For any positive number $\epsilon(9)$, there exists $T_{m,\epsilon(9),(6.42)}$ such that the next inequality holds for $T > T_{m,\epsilon(9),(6.42)}$.*

$$
\begin{aligned}
&\left\|\nabla_\rho^n \frac{d^\ell}{dT^\ell}\left(\mathfrak{P}^{-1}\left(\overline{\partial}h^0 - \mathcal{P}\left(\overline{\partial}h^1\right)\right)\right)\right\|_{L^2_{m-\ell,\delta}([4T-1,4T+1]_{\tau'}\times[0,1]\subset\Sigma_T)} \\
&\le \mu^\kappa\epsilon(9)e^{-\delta_1 T}.
\end{aligned}
\tag{6.42}
$$

PROOF. This is a consequence of 'drop of the weight' we mentioned in Remark 5.21.

The left hand side of (6.42) is the $L^2_{m-\ell,\delta}$ norm of the next formula

$$\nabla^n_\rho \frac{d^\ell}{dT^\ell}\Big(\mathfrak{P}^{-1}\left(\overline{\partial} h^0 - \mathcal{P}\left(\overline{\partial} h^1\right)\right)\Big) = -\int_0^1 \nabla^n_\rho \frac{d^\ell}{dT^\ell}\mathfrak{P}^{-1}\frac{\partial}{\partial r}\mathcal{P}(\overline{\partial} h^r)\,dr \tag{6.43}$$

Recalling the definition (6.41), we compute

$$\frac{\partial}{\partial r}\mathcal{P}\left(\overline{\partial} h^r\right) = \mathcal{P}D_{h^r}\overline{\partial}\left(\chi(\tau'-4T)\mathcal{P}^{-1}\left(V^\rho_{T,2,(\kappa)}(\tau'-10T,t) - (\Delta p^\rho_{T,(\kappa)})^{\mathrm{Pal}}\right)\right).$$

Therefore using the product rule (6.36) as before, we can estimate the integrand of (6.43)

$$\begin{aligned}
&\left\|\nabla^n_\rho \frac{d^\ell}{dT^\ell}\mathfrak{P}^{-1}\frac{\partial}{\partial r}\mathcal{P}(\overline{\partial} h^r)\right\|_{L^2_{m-\ell}} \\
\le & C_{m,(6.44)}\sum_{\ell'\le\ell}\sum_{n'\le n} \\
&\left\|\nabla^{n'}_\rho \frac{\partial^{\ell'}}{\partial T^{\ell'}}(\mathfrak{P}')^{-1}\Big(V^\rho_{T,2,(\kappa)}(\tau'',t)\right. \\
&\qquad\qquad \left.-(\Delta p^\rho_{T,(\kappa)})^{\mathrm{Pal}}\Big)\right\|_{L^2_{m+1-\ell'}([-6T-1,-6T+1]_{\tau''}\times[0,1])}.
\end{aligned} \tag{6.44}$$

in the way similar to the proof of (5.64) given at the end of Appendix A. Note the norm in (6.44) is the $L^2_{m+1-\ell'}$ norm without weight.

By the induction hypothesis the $L^2_{m+1-\ell'}$ norm of

$$\nabla^{n'}_\rho \frac{\partial^{\ell'}}{\partial T^{\ell'}}(\mathfrak{P}')^{-1}(V^\rho_{T,2,(\kappa)}(\tau'',t) - (\Delta p^\rho_{T,(\kappa)})^{\mathrm{Pal}})$$

with weight $e_{2,\delta}$ is estimated by $C_{2,m}\mu^{\kappa-1}e^{-\delta_1 T}$.

Over our domain $[4T-1,4T+1]_{\tau'}\times[0,1]$, the weight $e_{2,\delta}$ is around $e^{6T\delta}$. See Figure 5.1. Therefore

$$\left\|\nabla^n_\rho \frac{d^\ell}{dT^\ell}\mathfrak{P}^{-1}\Big(\int_0^1 \frac{\partial}{\partial r}\mathcal{P}(\overline{\partial} h^r)\,dr\Big)\right\|_{L^2_{m-\ell}} \le C_{m,(6.45)}e^{-6T\delta}\mu^{\kappa-1}e^{-\delta_1 T}. \tag{6.45}$$

On the other hand the weight $e_{1,\delta}$, which we use as the weight of the $L^2_{m-\ell,\delta}$ norm in the left hand side of (6.42) is around $e^{4T\delta}$.

Therefore

$$\text{LHS of (6.42)} \le C_{m,(6.46)}e^{-2T\delta}\mu^{\kappa-1}e^{-\delta_1 T}. \tag{6.46}$$

We take $T_{m,\epsilon(9),(6.42)} > 0$ such that $C_{m,(6.46)}e^{-2\delta T_{m,\epsilon(9),(6.42)}} \le \epsilon(9)\mu$. The lemma follows. □

Again applying (5.42) to the map $s \mapsto \overline{\partial}\left(\mathrm{Exp}(u^\rho_{T,(\kappa-1)}, sV^\rho_{T,1,(\kappa)}(\tau',t))\right)$ as in the proof of Proposition 5.34, we calculate

$$\begin{aligned}
&\overline{\partial}\left(\mathrm{Exp}(u^\rho_{T,(\kappa-1)}, V^\rho_{T,1,(\kappa)}(\tau',t))\right) \\
=&\overline{\partial} u^\rho_{T,(\kappa-1)}(\tau',t) + (D_{u^\rho_{T,(\kappa-1)}}\overline{\partial})(V^\rho_{T,1,(\kappa)})(\tau',t) \\
&+\int_0^1 ds\int_0^s \left(\frac{\partial^2}{\partial r^2}\right)\mathcal{P}\left(\overline{\partial}\left(\mathrm{Exp}(u^\rho_{T,(\kappa-1)}, rV^\rho_{T,1,(\kappa)}(\tau',t))\right)\right)dr.
\end{aligned} \tag{6.47}$$

Here $\mathcal{P}^{-1}$ is the complex linear part of the parallel transport along the path $s \mapsto \mathrm{Exp}(u^\rho_{T,(\kappa-1)}, sV^\rho_{T,1,(\kappa)}(\tau',t))$ $(s \in [0,r])$. We can estimate the third term of the right hand side of (6.47) in the same way as the proof of the inequality (G.3) given in Appendix G and obtain

$$\begin{aligned}&\left\|\nabla^n_\rho \frac{d^\ell}{dT^\ell}\mathfrak{P}^{-1}(\text{3rd term of (6.47)})\right\|_{L^2_{m-\ell}([4T-1,4T+1]_{\tau'}\times[0,1])}\\ &\le C_{m,(6.48)} \sum_{\ell'\le\ell, n'\le n}\left\|\nabla^{n'}_\rho \frac{\partial^{\ell'}}{\partial T^{\ell'}}(\mathfrak{P}')^{-1}V^\rho_{T,1,(\kappa)}\right\|^2_{L^2_{m+1-\ell'}([4T-1,4T+1]_{\tau'}\times[0,1])}.\end{aligned}\tag{6.48}$$

Since

$$\begin{aligned}&\left\|\nabla^{n'}_\rho \frac{\partial^{\ell'}}{\partial T^{\ell'}}(\mathfrak{P}')^{-1}V^\rho_{T,1,(\kappa)}\right\|_{L^2_{m+1-\ell'}([4T-1,4T+1]_{\tau'}\times[0,1])}\\ &\le \left\|\nabla^{n'}_\rho \frac{\partial^{\ell'}}{\partial T^{\ell'}}(\mathfrak{P}')^{-1}(V^\rho_{T,1,(\kappa)} - (\Delta p^\rho_{T,(\kappa)})^{\mathrm{Pal}})\right\|_{L^2_{m+1-\ell'}([4T-1,4T+1]_{\tau'}\times[0,1])}\\ &\quad + \left\|\nabla^{n'}_\rho \frac{\partial^{\ell'}}{\partial T^{\ell'}}(\mathfrak{P}')^{-1}(\Delta p^\rho_{T,(\kappa)})\right\|\\ &\le C_{m,(6.49)}\mu^{\kappa-1}e^{-\delta_1 T}\end{aligned}\tag{6.49}$$

by (6.9) we have

$$(6.48) \le C_{m,(6.50)}\mu^{2(\kappa-1)}e^{-2\delta_1 T}.\tag{6.50}$$

Finally we observe that

$$\overline{\partial}u^\rho_{T,(\kappa-1)} + (D_{u^\rho_{T,(\kappa-1)}}\overline{\partial})(V^\rho_{T,1,(\kappa)}(\tau',t)) = 0,\tag{6.51}$$

on $[4T-1,4T+1]_{\tau'}\times[0,1]$. This follows from (5.72) and Definition 5.31 (5.85) together with $\mathrm{Err}^\rho_{2,T,(\kappa)}(\tau',t) = 0$ on $[4T-1,4T+1]_{\tau'}\times[0,1]$.

In sum, by Lemma 6.16 and (6.47), (6.50) and (6.51), we obtain

$$\begin{aligned}&\left\|\nabla^n_\rho \frac{d^\ell}{dT^\ell}\mathfrak{P}^{-1}\mathcal{P}(\mathrm{Err}^\rho_{1,T,(\kappa)})\right\|_{L^2_{m-\ell,\delta}([4T-1,4T+1]_{\tau'}\times[0,1])}\\ &\le C_{m,(6.52)}\mu^\kappa\epsilon(9)e^{-\delta_1 T},\end{aligned}\tag{6.52}$$

for $T > T_{m,\epsilon(9),(6.52)}$.

It implies

$$\begin{aligned}&\left\|\nabla^n_\rho \frac{d^\ell}{dT^\ell}\mathfrak{P}^{-1}(\mathrm{Err}^\rho_{1,T,(\kappa)})\right\|_{L^2_{m-\ell,\delta}([4T-1,4T+1]_{\tau'}\times[0,1])}\\ &\le C_{m,(6.53)}\mu^\kappa\epsilon(9)e^{-\delta_1 T}.\end{aligned}\tag{6.53}$$

In fact $\mathfrak{P}^{-1}\mathcal{P}$ in (6.52) is induced by $\mathrm{Pal}^{u_1}_{h^0}\circ\mathrm{Pal}^{h^0}_{h^1}$ and $\mathfrak{P}^{-1}$ in (6.53) is induced by $\mathrm{Pal}^{u_1}_{h^1}$. Therefore we can show that (6.52) implies (6.53) using (6.9)-(6.11) and Lemma 6.12 for κ, $\kappa-1$. Namely we can use them to estimate T and ρ derivatives of

$$\mathrm{Pal}^{u_1}_{h^1}\circ\mathrm{Pal}^{h^1}_{h^0}\circ\mathrm{Pal}^{h^0}_{u_1}$$

in the same way as in Appendix B.

We can now complete the proof of Proposition 6.9 (1). We take $\epsilon(7)$, $\epsilon(8)$, $\epsilon(9)$ such that $\epsilon(7) < C_{8,m}\epsilon(6)/10$, $C_{m,(6.39)}\epsilon(8) < C_{8,m}\epsilon(6)/10$ and $\epsilon(9) < C_{8,m}\epsilon(6)/10$. Then if $T > \max\{T_{m,\epsilon(7),(6.25)}, T_{m,\epsilon(8),(6.34)}, T_{m,\epsilon(9),(6.42)}, T_{m,\epsilon(9),(6.52)}\}$ Lemmata 6.14, 6.16 and Formulae (6.39), (6.53) imply (6.12). (6.13) then follows from Lemma 6.14 (6.26). □

REMARK 6.17. In [**Ab**] Abouzaid used the L^p_1 norm for the maps u. He then proved that the gluing map is continuous with respect to T (that is S in the notation of [**Ab**]) but does not prove its differentiability with respect to T. (Instead he used the technique to remove the part of the moduli space with $T > T_0$. This technique certainly works for the purpose of [**Ab**].) In fact if we use the L^p_1 norm instead of the L^2_m norm then the left hand side of (6.11) becomes the L^p_{-1} norm which is hard to use.

Abouzaid mentioned in [**Ab**, Remark 5.1] that this point is related to the fact that quotients of Sobolev spaces by the diffeomorphisms in the source are not naturally equipped with the structure of smooth Banach manifold. Indeed in the situation where there is an automorphism on Σ_2, for example when Σ_2 is the disk with one boundary marked point at ∞, then the T parameter is killed by a part of the automorphism. So the shift of $V^\rho_{T,2,(\kappa)}$ by T that appears in the second term of (6.21) will be equivalent to the action of the automorphism group of Σ_2 in such a situation. The shift of T causes the loss of differentiability in the sense of Sobolev space in the formulas (6.9) -(6.13). However at the end of the day we can still get the differentiability of C^∞ order and its exponential decay by using various *weighted* Sobolev spaces with various m simultaneously using the fact that C^∞ topology is a Frechêt topology, as we show during the proof of Lemma 8.28. (See Remark 6.3 also.)

REMARK 6.18. In [**DK**, Subsection 7.2.3] Donaldson-Kronheimer mentioned that there is exactly one place where their construction of the basic package on the moduli space of ASD-connections uses the L^p_m space for $p \neq 2$ which is not conformally invariant. It is exactly the place of gluing construction, the place similar to what we are studying in this paper. There Donaldson-Kronheimer used the L^p_1 space. The reason they do need the L^p_m norm for $p \neq 2$ is that in [**DK**] Donaldson-Kronheimer do not use *weighted* Sobolev or Banach norm.

In the framework of [**DK**] (which is adapted to the pseudoholomorphic curve by [**McSa**]) the 'neck region' of 4 manifold is regarded as $D^4(1) \setminus D^4(1/R^2)$ with the standard Riemannian metric on $D^4(1) \setminus D^4(1/R)$ and the metric induced from the standard metric by $x \mapsto Rx/|x|^2$ on $D^4(1/R) \setminus D^4(1/R^2)$. (In [**McSa**] the 'neck region' of the source curve is $D^2(1) \setminus D^2(1/R^2)$ with a similar metric.) So their metric is different from the cylindrical metric on $[-5T, 5T] \times S^1$, which is one we use in this paper.

If we change the variables from $D^2(1) \setminus D^2(1/R^2)$ to $[-5T, 5T] \times S^1$ (with $R = e^{5\pi T}$) but still use the above mentioned Riemannian metric of $D^2(1) \setminus D^2(1/R^2)$ then it is equivalent to using the cylindrical metric of $[-5T, 5T] \times S^1$ together with some weight. Note this weight function is 1 if and only if the Sobolev space involved is conformally invariant. So in the situation such as the one appearing in [**DK**, Subsection 7.2.3] this weight function is nontrivial. Actually one can observe that the weight function appears in that way is similar to the weight $e_{\delta,T}$, which we use in this paper. In other words, if we use an appropriate L^p_1 norm, using the cylindrical metric with weight is not very different from using the the above mentioned metric on $D^2(1) \setminus D^2(1/R^2)$.

However if we consider the L^2_m norm with m large enough then our weighted norm (after changing the variables to $D^2(1) \setminus D^2(1/R^2)$) does not coincide with the L^2_m norm with respect to the above mentioned metric. Thus it seems important to use a weighted norm for the study of higher derivatives with respect to the gluing parameter T.

We remark that in [**FU**] Freed-Uhlenbeck worked out the gluing analysis of ASD connections in the frame work of L^2 theory (that is, without using L^p_k spaces but using only L^2_k spaces). Freed-Uhlenbeck used the cylindrical metric on $\mathbb{R} \times S^3$. So the method of [**FU**] is closer to ours. It seems that Freed-Uhlenbeck do not need to use weighted Sobolev norm since in their case they can use the fact that their 3 manifold is S^3 and show exponential decay without using weighted Sobolev norm. Actually, in their situation, the Chern-Simons functional on S^3 is not only a Morse-Bott function but also a Morse function. In our situation, the non-linear Cauchy-Riemann equation on $\mathbb{R} \times S^1$ or on $\mathbb{R} \times [0,1]$ with Lagrangian boundary condition is degenerate at infinity. In other words we are in Morse-Bott situation. By this reason, it seems inevitable to use weighted Sobolev norm when we work with the cylindrical metric.

REMARK 6.19. Another difference between our construction and the construction of [**DK**],[**McSa**] is the choice of cut off function. In the formula at the end of [**McSa**, page 172] they used the cut off function β appearing in [**DK**, Lemma 7.2.10], [**McSa**, Lemma A.A.1] to obtain the right inverse. If we rewrite their formula in terms of the cylindrical coordinates the cut off function appearing there has mostly of constant slope $|1/\log \delta|$ and the support of its first derivative has length $\sim |\log \delta|$, here δ is a small number. So the size of the error term caused by the derivative of the cut off function is (pointwise) $\sim |1/\log \delta|$, which is small.

Our choice of cut off function is, for example, $\chi^{\leftarrow}_{\mathcal{A}}$. In the cylindrical coordinates the size of its derivative is ~ 1 and the length of the support of its first derivative is also ~ 1. So the size of the error term caused by the derivative of the cut off function is (pointwise) ~ 1, which is *not* small. We use the 'drop of the weight argument' mentioned in Remark 5.21 to show that this error term is small in our *weighted* Sobolev norm. See Figure 6.3.

Since we use the *weighted* Sobolev space, the estimate is easier to carry out in case the support of the derivative of the cut off function has bounded length. (This is because then the ratio between maximum and minimum of the weight function on the above mentioned support is bounded.)

6.3. Part B: Estimates for the approximate inverse

In this section we prove Proposition 6.9 (2). This section corresponds to the discussion given in [**FOOO1**, page 776 the paragraph next to (B)].

We assume (6.9)-(6.13) for κ and will prove (6.9) for $\kappa + 1$. (So we are doing Step $\kappa + 1$.)

This part is nontrivial only because the construction here is global. (Solving linear equation.) So we first review the set-up of the function space that is independent of T, ρ, κ.

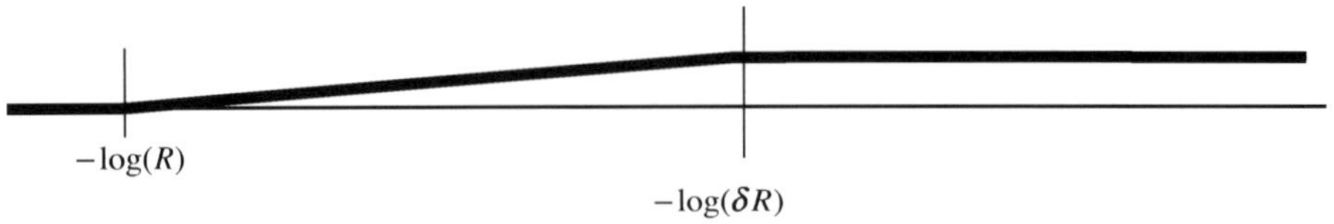

[DK]'s choice of cut off function

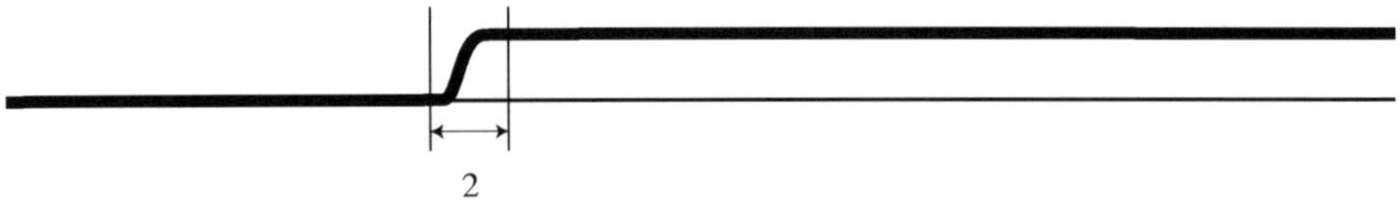

the choice of cut off function in this paper

FIGURE 6.3. The choices of cut off functions.

In Definition 5.30 we defined a function space $\mathfrak{H}_{(\kappa)}(\mathcal{E}_1, \mathcal{E}_2; \rho, T)$, which is a subspace of (5.84). We solved the linearized equation on it. (See (5.86).) The space (5.84) is T, ρ, κ-dependent. However $\mathfrak{H}_{(\kappa)}(\mathcal{E}_1, \mathcal{E}_2; \rho, T)$ is the image of $\mathfrak{H}(\mathcal{E}_1, \mathcal{E}_2)$, which is independent of T, ρ, κ, by the map $(\Phi_{1;(\kappa)}(\rho, T), \Phi_{2;(\kappa)}(\rho, T))$. We recall $\mathfrak{H}(\mathcal{E}_1, \mathcal{E}_2)$ is defined in Definition 5.9. We put

$$I^0_{(\kappa),\rho,T} = (\Phi_{1;(\kappa)}(\rho, T), \Phi_{2;(\kappa)}(\rho, T)) : \mathfrak{H}(\mathcal{E}_1, \mathcal{E}_2) \to \mathfrak{H}_{(\kappa)}(\mathcal{E}_1, \mathcal{E}_2; \rho, T).$$

By composing it we can take the domain independent of T, ρ, κ.

We next consider the codomain of the above approximate linearization. Using the map in (2.7) we define

$$I^1_{i,(\kappa),\rho,T} : L^2_{m,\delta}(\Sigma_i; u_i^*TX \otimes \Lambda^{0,1}) \to L^2_{m,\delta}(\Sigma_i; (\hat{u}^\rho_{i,T,(\kappa)})^*TX \otimes \Lambda^{0,1})$$

as the closure of $(\mathrm{Pal}_{u_i}^{\hat{u}^\rho_{i,T,(\kappa)}})^{(0,1)}$.

Then we define

$$\begin{aligned} I^1_{(\kappa),\rho,T} : \; & L^2_{m,\delta}(\Sigma_1; u_1^*TX \otimes \Lambda^{0,1}) \oplus L^2_{m,\delta}(\Sigma_2; u_2^*TX \otimes \Lambda^{0,1}) \\ & \longrightarrow L^2_{m,\delta}(\Sigma_1; (\hat{u}^\rho_{1,T,(\kappa)})^*TX \otimes \Lambda^{0,1}) \oplus L^2_{m,\delta}(\Sigma_2; (\hat{u}^\rho_{2,T,(\kappa)})^*TX \otimes \Lambda^{0,1}) \end{aligned}$$

as the direct sum $I^1_{(\kappa),\rho,T} = I^1_{1;(\kappa),\rho,T} \oplus I^1_{2;(\kappa),T}$. Thus the composition

$$(I^1_{(\kappa),\rho,T})^{-1} \circ \left(D_{\hat{u}^\rho_{1,T,(\kappa)}} \overline{\partial} \oplus D_{\hat{u}^\rho_{2,T,(\kappa)}} \overline{\partial} \right) \circ I^0_{(\kappa),\rho,T}$$

defines an operator, which we denote by

$$D_{(\kappa),\rho,T} : \mathfrak{H}(\mathcal{E}_1, \mathcal{E}_2) \to L^2_{m,\delta}(\Sigma_1; u_1^*TX \otimes \Lambda^{0,1}) \oplus L^2_{m,\delta}(\Sigma_2; u_2^*TX \otimes \Lambda^{0,1}). \tag{6.54}$$

Both domain and codomain of this operator now are independent of κ, ρ, T.

We need to invert the operator $D^{\mathrm{app},(\kappa)}_{\hat{u}^\rho_{1,T,(\kappa)}} \overline{\partial} \oplus D^{\mathrm{app},(\kappa)}_{\hat{u}^\rho_{2,T,(\kappa)}} \overline{\partial}$ modulo $\mathcal{E}_1(\hat{u}^\rho_{1,T,(\kappa)}) \oplus \mathcal{E}_2(\hat{u}^\rho_{2,T,(\kappa)})$. (See (5.30) and (5.85).) We remark the subspace

$$\mathcal{E}_{i,(\kappa),\rho,T} := (I^1_{(\kappa),\rho,T})^{-1}(\mathcal{E}_1(\hat{u}^\rho_{i,T,(\kappa)}) \oplus \mathcal{E}_2(\hat{u}^\rho_{i,T,(\kappa)})) \tag{6.55}$$

is different from $\mathcal{E}_1(u_1)\oplus\mathcal{E}_2(u_2)$ and is (κ,ρ,T)-dependent in general: In fact, by definition $\mathcal{E}_i(\hat{u}^\rho_{i,T,(\kappa)})$ is the image of $\mathcal{E}_i^{\mathfrak{ob}}$ under the parallel transport $(\mathrm{Pal}_{u_i^{\mathfrak{ob}}}^{\hat{u}^\rho_{i,T,(\kappa)}})^{(0,1)}$. (See (3.18).) Therefore

$$(I^1_{(\kappa),\rho,T})^{-1}(\mathcal{E}_1(\hat{u}^\rho_{1,T,(\kappa)})\oplus\mathcal{E}_2(\hat{u}^\rho_{2,T,(\kappa)}))$$
$$=\bigoplus_{i=1}^2\left(\left((\mathrm{Pal}_{u_i}^{\hat{u}^\rho_{i,T,(\kappa)}})^{(0,1)}\right)^{-1}\circ(\mathrm{Pal}_{u_i^{\mathfrak{ob}}}^{\hat{u}^\rho_{i,T,(\kappa)}})^{(0,1)}\right)(\mathcal{E}_i^{\mathfrak{ob}})$$

which is different from $\mathcal{E}_1(u_1)\oplus\mathcal{E}_2(u_2)$ since

$$(\mathrm{Pal}_{u_i^{\mathfrak{ob}}}^{u_i})^{(0,1)}(\mathcal{E}_i^{\mathfrak{ob}})\neq\left(\left((\mathrm{Pal}_{u_i}^{\hat{u}^\rho_{i,T,(\kappa)}})^{(0,1)}\right)^{-1}\circ(\mathrm{Pal}_{u_i^{\mathfrak{ob}}}^{\hat{u}^\rho_{i,T,(\kappa)}})^{(0,1)}\right)(\mathcal{E}_i^{\mathfrak{ob}}) \tag{6.56}$$

in general.

REMARK 6.20. In our situation where we consider only one $u_i^{\mathfrak{ob}}$ we can trivialize the bundle $u'\mapsto L^2_m(\Sigma_T,(u')^*TX\otimes\Lambda^{0,1})$ over $\{u'\}$ near $\hat{u}^\rho_{T,(\kappa)}$ by sending them to

$$\bigoplus_{i=1}^2 L^2_m(\Sigma_i,(u_i^{\mathfrak{ob}})^*TX\otimes\Lambda^{0,1}) \tag{6.57}$$

using parallel transport. Then the image of $\mathcal{E}_i(u')$ in (6.57) will not vary.

However for our application, we need to consider several different $u_i^{\mathfrak{ob}}$'s, in which case there is no choice of the trivialization of $u'\mapsto\mathcal{E}_i(u')\subset L^2_m(\Sigma_T,(u')^*TX\otimes\Lambda^{0,1})$ so that the direct sum of the obstruction spaces do not vary.

However the way to estimate this discrepancy in the case when we have several $u_i^{\mathfrak{ob}}$'s is the same as we do in this section.

We can estimate the (T,ρ) and κ dependence of (6.55) and use certain elementary functional analysis to go around the problem of this dependence as follows.

We first observe the next lemma. Let $\{\mathbf{e}_{i,a}\mid a=1,\dots,\dim\mathcal{E}_i^{\mathfrak{ob}}\}$ be a basis of $\mathcal{E}_i^{\mathfrak{ob}}$. We put

$$\begin{aligned}\mathbf{e}'_{i,a;(\kappa)}(\rho,T)&=\left(\left((\mathrm{Pal}_{u_i}^{\hat{u}^\rho_{i,T,(\kappa)}})^{(0,1)}\right)^{-1}\circ(\mathrm{Pal}_{u_i^{\mathfrak{ob}}}^{\hat{u}^\rho_{i,T,(\kappa)}})^{(0,1)}\right)(\mathbf{e}_{i,a})\\&\in L^2_{m,\delta}(\Sigma_i;u_i^*TX\otimes\Lambda^{0,1}).\end{aligned} \tag{6.58}$$

Note $\{\mathbf{e}'_{i,a;(\kappa)}(\rho,T)\mid a=1,\dots,\dim\mathcal{E}_i^{\mathfrak{ob}}\}$ is a basis of $\mathcal{E}_{i,(\kappa),\rho,T}$.

We denote by $\mathbf{e}_{i,a;(\kappa)}(\rho,T)$ $(a=1,\dots,\dim\mathcal{E}_i^{\mathfrak{ob}})$ the basis obtained from

$$\mathbf{e}'_{i,a;(\kappa)}(\rho,T)$$

by applying the Gram-Schmidt orthogonalization to $\mathbf{e}'_{i,a;(\kappa)}(\rho,T)$. (We use the L^2 inner product (4.11) on $L^2_{m,\delta}(\Sigma_i;u_i^*TX\otimes\Lambda^{0,1})$.)

LEMMA 6.21. *There exists $C_{m,(6.59)}$ such that*

$$\left\|\nabla_\rho^n\frac{\partial^\ell}{\partial T^\ell}\mathbf{e}_{i,a;(\kappa)}(\rho,T)\right\|_{L^2_{m+1-\ell}}\leq C_{m,(6.59)}e^{-\delta_1 T}, \tag{6.59}$$

for $m-2\geq n\geq 0$, $m-2\geq\ell>0$.

We remark that since the support of $\mathbf{e}_{i,a;(\kappa)}(\rho,T)$ is in K_i we do not need to use a weighted norm.

PROOF. We prove this lemma in Appendix F. □

Let $\mathcal{E}^{\perp}_{i,(\kappa),\rho,T}$ be the L^2 orthonormal complement of $\mathcal{E}_{i,(\kappa),\rho,T}$ in $L^2_{m,\delta}(\Sigma_i; u_i^*TX\otimes\Lambda^{0,1})$. In other words, we define

$$\mathcal{E}^{\perp}_{i,(\kappa),\rho,T} = \{W \in L^2_{m,\delta}(\Sigma_i; u_i^*TX \otimes \Lambda^{0,1}) \mid \langle\langle W, \mathbf{e}_{i,a;(\kappa)}(\rho,T)\rangle\rangle_{L^2} = 0,\ a = 1,\dots,\dim \mathcal{E}_i^{\mathfrak{ob}}\}.$$

Here we use the L^2 inner product defined in (4.11). Since elements $\mathbf{e}_{i,a;(\kappa)}(\rho,T)$ are supported in K_i and are smooth, we can safely use the L^2 inner product, without weight. We also define the projections

$$\Pi_{\mathcal{E}_{i,(\kappa),\rho,T}} : L^2_{m,\delta}(\Sigma_i; u_i^*TX \otimes \Lambda^{0,1}) \to \mathcal{E}_{i,(\kappa),\rho,T},$$

by

$$\Pi_{\mathcal{E}_{i,(\kappa),\rho,T}}(W) = \sum_{a=1}^{\dim \mathcal{E}_i^{\mathfrak{ob}}} \langle\langle W, \mathbf{e}_{i,a,(\kappa)}(\rho,T)\rangle\rangle_{L^2}\mathbf{e}_{i,a,(\kappa)}(\rho,T). \tag{6.60}$$

We put

$$\mathcal{E}_{(\kappa),\rho,T} = \bigoplus_{i=1}^{2} \mathcal{E}_{i,(\kappa),\rho,T}, \quad \mathcal{E}^{\perp}_{(\kappa),\rho,T} = \bigoplus_{i=1}^{2} \mathcal{E}^{\perp}_{i,(\kappa),\rho,T}$$

and $\Pi_{\mathcal{E}_{(\kappa),\rho,T}} = \Pi_{\mathcal{E}_{1,(\kappa),\rho,T}} \oplus \Pi_{\mathcal{E}_{2,(\kappa),\rho,T}}$. The operator we need to invert is

$$(\mathrm{id} - \Pi_{\mathcal{E}_{(\kappa),\rho,T}}) \circ D_{(\kappa),\rho,T} : \mathfrak{H}(\mathcal{E}_1,\mathcal{E}_2) \to \mathcal{E}^{\perp}_{(\kappa),\rho,T}.$$

Let $\mathcal{E}_i^{\perp}$ be the L^2 orthogonal complement of $\mathcal{E}_i = \mathcal{E}_i(u_i)$ in $L^2_{m,\delta}(\Sigma_i; u_i^*TX \otimes \Lambda^{0,1})$, $\Pi^{\perp}_{\mathcal{E}_i} : L^2_{m,\delta}(\Sigma_i; u_i^*TX \otimes \Lambda^{0,1}) \to \mathcal{E}_i^{\perp}$ the associated L^2 projection. We put $\Pi = \Pi^{\perp}_{\mathcal{E}_1} \oplus \Pi^{\perp}_{\mathcal{E}_2}$. By Lemma 6.21 the restriction of Π induces an isomorphism

$$\Pi : \mathcal{E}^{\perp}_{(\kappa),\rho,T} \to \mathcal{E}_1^{\perp} \oplus \mathcal{E}_2^{\perp}. \tag{6.61}$$

The map $D_{(\kappa),\rho,T}$ induces a map

$$\overline{D}_{(\kappa),\rho,T} = \Pi \circ (\mathrm{id} - \Pi_{\mathcal{E}_{(\kappa),\rho,T}}) \circ D_{(\kappa),\rho,T} : \mathfrak{H}(\mathcal{E}_1,\mathcal{E}_2) \to \mathcal{E}_1^{\perp} \oplus \mathcal{E}_2^{\perp}. \tag{6.62}$$

Both the domain and the codomain of $\overline{D}_{(\kappa),\rho,T}$ are independent of κ, T, ρ. We will invert this operator and examine its T, ρ dependence by using the next lemma.

Recall from Definition 5.9 that $\mathfrak{H}(\mathcal{E}_1,\mathcal{E}_2)$ is a subspace of

$$\bigoplus_{i=1}^{2} W^2_{m+1,\delta}((\Sigma_i,\partial\Sigma_i); u_i^*TX, u_i^*TL),$$

consisting of the pairs (V_1,V_2) where $V_i = (s_i,v_i)$ is a pair of a section s_i and its asymptotic value v_i. (Note $v_1 = v_2$ for an element of $\mathfrak{H}(\mathcal{E}_1,\mathcal{E}_2)$.)

LEMMA 6.22. *For $V \in \mathfrak{H}(\mathcal{E}_1,\mathcal{E}_2)$ the following holds:*

(1) *There exist $C_{m,(6.63)}, C'_{m,(6.63)} > 0$ such that*

$$C_{m,(6.63)}\|V\|_{W^2_{m+1,\delta}} \le \|\overline{D}_{(\kappa),\rho,T}(V)\|_{L^2_{m,\delta}} \le C'_{m,(6.63)}\|V\|_{W^2_{m+1,\delta}}. \tag{6.63}$$

(2) *There exist* $C_{m,(6.64)}, C_{m,(6.65)} > 0$ *such that*

$$\left\|\nabla_\rho^n(\overline{D}_{(\kappa),\rho,T}(V) - \overline{D}_{(0),\rho,T}(V))\right\|_{L^2_{m,\delta}} \le C_{m,(6.64)} e^{-\delta_1 T/10} \|V\|_{W^2_{m+1,\delta}}, \tag{6.64}$$

for $m \ge n$ *and*

$$\left\|\nabla_\rho^n \frac{\partial^\ell}{\partial T^\ell} \overline{D}_{(\kappa),\rho,T}(V)\right\|_{L^2_{m-\ell,\delta}} \le C_{m,(6.65)} e^{-\delta_1 T/10} \|V\|_{W^2_{m+1,\delta}} \tag{6.65}$$

for $m \ge n \ge 0, m \ge \ell > 0$.

PROOF. The inequality (6.63) follows from (6.64) and the invertibility of the operator $\overline{D}_{(0),\rho,T}$.

The proofs of (6.64), (6.65) will occupy the rest of the proof.

First, we remark that, by Lemma 6.21 and (6.60) we have:

$$\left\|\nabla_\rho^n \frac{\partial^\ell}{\partial T^\ell} (\Pi \circ (\mathrm{id} - \Pi_{\mathcal{E}_{(\kappa),\rho,T}}))\right\|_{L^2_{m+1-\ell}} \le C_{m,(6.66)} e^{-\delta_1 T}. \tag{6.66}$$

We next study $D_{(\kappa),\rho,T}$. By (6.10) and (6.11) (See also Lemma 6.12.), we obtain the next inequalities by induction hypothesis:

$$\left\|\nabla_\rho^n \mathscr{P}_{i,\rho_i} \mathrm{E}(u_i^\rho, \hat{u}^\rho_{i,T,(\kappa)})\right\|_{L^2_{m+1-\ell}(K_i^T)} \le C_{m,(6.67)} e^{-\delta_1 T}, \tag{6.67}$$

$$\left\|\nabla_\rho^n \mathrm{Pal}^{p_0}_{p_0^\rho} (\mathrm{E}(p_0^\rho, \hat{u}^\rho_{i,T,(\kappa)}))\right\|_{L^2_{m+1-\ell}(K_i^{9T} \setminus K_i^T)} \le C_{m,(6.68)} e^{-\delta_1 T}, \tag{6.68}$$

and

$$\left\|\nabla_\rho^n \frac{\partial^\ell}{\partial T^\ell} \mathrm{E}(u_i, \hat{u}^\rho_{i,T,(\kappa)})\right\|_{L^2_{m+1-\ell}(K_i^T)} \le C_{m,(6.69)} e^{-\delta_1 T}, \tag{6.69}$$

$$\left\|\nabla_\rho^n \frac{\partial^\ell}{\partial T^\ell} \mathrm{Pal}^{p_0}_{p_0^\rho} (\mathrm{E}(p_0^\rho, \hat{u}^\rho_{i,T,(\kappa)}))\right\|_{L^2_{m+1-\ell}(K_i^{9T} \setminus K_i^T)} \le C_{m,(6.70)} e^{-\delta_1 T}. \tag{6.70}$$

Here $m - 2 \ge n$ and $m - 2 \ge \ell > 0$. Note we use τ' as the coordinate of $[0,\infty)_{\tau'}$ and τ'' as the coordinate of $(-\infty, 0]_{\tau''}$ in the left hand sides of (6.68) and (6.70), respectively. Hereafter we assume n, ℓ satisfy $m - 2 \ge n$ and $m - 2 \ge \ell \ge 0$ until the end of the proof of Lemma 6.22. In fact (6.68), (6.70) follow from (6.11). The inequality (6.67) follows from (6.10). The inequality (6.69) follows from (6.10).

We also remark that

$$\hat{u}^\rho_{i,T,(\kappa-1)} \equiv p^\rho_{T,(\kappa-1)}$$

on $\Sigma_i \setminus K_i^{7T+1}$.

We will derive (6.64), (6.65) with $\overline{D}_{(\kappa),\rho,T}$ replaced by $D_{(\kappa),\rho,T}$ from these four inequalities as follows.

We put $V = (V_1, V_2)$ and $V_i = (s_i, v_i)$.

We consider the case of Σ_1 only since the case of Σ_2 is the same. We divide the domain Σ_1 into the pieces K_1 and $[k-1, k+1]_{\tau'} \times [0,1]$, ($k \in [0,\infty) \cap \mathbb{Z}$).

We denote each piece of these domains by $\Sigma(a)$. For each such piece $\Sigma(a)$, we define $\Sigma(a,+)$ as follows. (We use the τ' coordinate.)

(a) If $\Sigma(a) = K_1$ then $\Sigma(a,+) = K_1 \cup [0,1]_{\tau'} \times [0,1]$.
(b) If $\Sigma(a) = [k-1, k+1]_{\tau'} \times [0,1]$ then $\Sigma(a,+) = [k-2, k+2]_{\tau'} \times [0,1]$.

We recall from the definition (6.62) and (6.54) that $D_{(0),\rho,T}$, $\overline{D}_{(0),\rho,T}$ is independent of T for any $\rho \in \overline{V_1(\epsilon_2) \times_L V_2(\epsilon_2)}$. So we denote the common operators by D_ρ, $\overline{D}_\rho$.

We first consider the case when a is as in (a) above. Then from (6.67),(6.69) we derive

$$\begin{aligned}&\left\|\nabla_\rho^n \frac{\partial^\ell}{\partial T^\ell}(D_{(\kappa),\rho,T} - D_\rho)(V)\right\|_{L^2_{m-\ell}(\Sigma(a))}\\ &\le C_{m,(6.71)} e^{-\delta_1 T} \|s_1\|_{L^2_{m+1}(\Sigma(a,+))}.\end{aligned} \tag{6.71}$$

(Here we use the fact that V is independent of T, ρ.)

We next consider the case when a is as in (b), that is, $\Sigma(a) = [k-1,k+1]_{\tau'} \times [0,1]$. Then

$$\begin{aligned}&\left\|\nabla_\rho^n \frac{\partial^\ell}{\partial T^\ell}(D_{(\kappa),\rho,T} - D_\rho)(V)\right\|_{L^2_{m-\ell}(\Sigma(a))}\\ &\le \left\|\nabla_\rho^n \frac{\partial^\ell}{\partial T^\ell}(D_{(\kappa),\rho,T} - D_\rho)(s_1 - v_1^{\rm pal})\right\|_{L^2_{m-\ell}(\Sigma(a))}\\ &\quad + \left\|\nabla_\rho^n \frac{\partial^\ell}{\partial T^\ell}(D_{(\kappa),\rho,T} - D_\rho)(v_1^{\rm pal})\right\|_{L^2_{m-\ell}(\Sigma(a))} \qquad (6.72)\\ &\le C_{m,(6.73)} e^{-\delta_1 T} \|s_1 - v_1^{\rm pal}\|_{L^2_{m+1}(\Sigma(a,+))} + C_{m,(6.73)} e^{-\delta_1 T}\|v_1\|. \qquad (6.73)\end{aligned}$$

Here we use (6.9)-(6.13) to derive the inequality (6.73).

Moreover for those τ' with $\Sigma(a,+) \cap [0, 7T+1]_{\tau'} \times [0,1] = \emptyset$ we may improve (6.73) to be $\le C_{m,(6.73)} e^{-\delta_1} \|s_1 - v_1^{\rm pal}\|_{L^2_m(\Sigma(a,+))}$. (Namely the second term of (6.73) drops out.) This is because $\hat u^\rho_{1,T,(\kappa)}$ is then constant on $\Sigma(a,+)$ and therefore $D_{(\kappa),\rho,T}(v_1^{\rm pal}) = 0$ thereon.

The norm

$$\sum_a e_{1,\delta}(p(a))^2 \|Y\|^2_{L^2_{m+1}(\Sigma(a))},$$

is equivalent to the $L^2_{m+1,\delta}$ norm $\|Y\|^2_{L^2_{m+1,\delta}(\Sigma_1)}$. (Here $e_{1,\delta}$ is the weight function in (4.4) and $p(a) \in \Sigma(a)$ is any chosen point for each a.)

Moreover the norm

$$\sum_{a:\text{Case (a)}} \|s_1\|^2_{L^2_{m+1}(\Sigma(a))} + \sum_{a:\text{Case (b)}} e_{1,\delta}(p(a))^2 \left\|s_1 - v_1^{\rm Pal}\right\|^2_{L^2_{m+1}(\Sigma(a))} + \|v_1\|^2,$$

is equivalent to the $W^2_{m+1,\delta}$ norm $\|(s_1,v_1)\|^2_{W^2_{m+1,\delta}(\Sigma_1)}$.

Therefore taking the weighted sum of the square of the inequalities (6.71), (6.73) with weight $e_{T,\delta}(p(a))^2$ and using the above mentioned equivalence of norms, we derive the estimate

$$
\begin{aligned}
&\left\|\nabla_\rho^n \frac{\partial^\ell}{\partial T^\ell}(D_{(\kappa),\rho,T} - D_\rho)(V)\right\|^2_{L^2_{m-\ell,\delta}(\Sigma_1)} \\
&\le C_{m,(6.74)} e^{-2\delta_1 T} \sum_{a:\text{Case (a)}} \|s_1\|^2_{L^2_{m+1}(\Sigma(a,+))} \\
&+ C'_{m,(6.74)} \sum_{a:\text{Case (b)}} e^{-2\delta_1 T} e_{1,\delta}(p(a))^2 \left\|s_1 - v_1^{\mathrm{Pal}}\right\|^2_{L^2_{m+1}(\Sigma(a,+))} \\
&+ T e^{-2\delta_1 T} e^{14\delta T} C''_{m,(6.74)} \|v_1\|^2.
\end{aligned} \tag{6.74}
$$

Here we use the fact that if $\Sigma(a,+)\cap[0,7T+1]_{\tau'}\times[0,1]\neq\emptyset$ then $e_{1,\delta}(p(a))\le Ce^{7\delta T}$. Using $\delta<\delta_1/10$ from (3.11), (6.74) implies

$$
\begin{aligned}
&\left\|\nabla_\rho^n \frac{\partial^\ell}{\partial T^\ell}(D_{(\kappa),\rho,T} - D_\rho)(V)\right\|_{L^2_{m-\ell,\delta}(\Sigma_1)} \\
&\le C_{m,(6.75)} e^{-\delta_1 T/10} \|(s_1,v_1)\|_{W^2_{m+1,\delta}(\Sigma_1)}.
\end{aligned} \tag{6.75}
$$

This inequality and the same inequality for Σ_2 together with (6.62) and (6.66) imply (6.65). □

Since the operator $\overline{D}_{(\kappa),\rho,T}$ is invertible, we can expand the operator

$$
\overline{D}^{-1}_{(\kappa),\rho,T} : \mathcal{E}_1^\perp \oplus \mathcal{E}_2^\perp \to \mathfrak{H}(\mathcal{E}_1,\mathcal{E}_2)
$$

into

$$
\begin{aligned}
\overline{D}^{-1}_{(\kappa),\rho,T} &= \left((1 + (\overline{D}_{(\kappa),\rho,T} - \overline{D}_\rho)\overline{D}_\rho^{-1})\overline{D}_\rho\right)^{-1} \\
&= \overline{D}_\rho^{-1} \sum_{k=0}^{\infty} (-1)^k ((\overline{D}_{(\kappa),\rho,T} - \overline{D}_\rho)\overline{D}_\rho^{-1})^k.
\end{aligned} \tag{6.76}
$$

It follows from (6.63), (6.64) that the right hand side converges as an operator from $(\mathcal{E}_1^\perp \oplus \mathcal{E}_2^\perp)\cap L^2_{m,\delta}$ to $W^2_{m+1,\delta}$ as far as ρ is in a sufficiently small neighborhood $V(\rho'_0)$ of ρ'_0 and $T > T_{m,(6.76)}$

Therefore by differentiating this by T using the Leibnitz rule, we derive

$$
\begin{aligned}
\frac{\partial}{\partial T}\overline{D}^{-1}_{(\kappa),\rho,T} = \overline{D}_\rho^{-1} &\sum_{k_1=0}^{\infty}\sum_{k_2=0}^{\infty} (-1)^{k_1+k_2+1} \left((\overline{D}_{(\kappa),\rho,T} - \overline{D}_\rho)\overline{D}_\rho^{-1}\right)^{k_1} \\
&\times \left(\frac{\partial}{\partial T}\overline{D}_{(\kappa),\rho,T}\overline{D}_\rho^{-1}\right)\left((\overline{D}_{(\kappa),\rho,T} - \overline{D}_\rho)\overline{D}_\rho^{-1}\right)^{k_2}.
\end{aligned}
$$

Along the way, we lose one degree of differentiability. Here we note that the operator $(\overline{D}_{(\kappa),\rho,T} - \overline{D}_\rho)\overline{D}_\rho^{-1}$ is uniformly bounded as κ,ρ,T vary with $\rho\in V(\rho'_0)$, $T > T_{m,(6.76)}$. In the same way, we can differentiate with respect to ρ without loss of derivative.

In a similar way we can derive

$$
\left\|\nabla_\rho^n \frac{\partial^\ell}{\partial T^\ell}\overline{D}^{-1}_{(\kappa),\rho,T}(W)\right\|_{W^2_{m+1-\ell,\delta}} \le C_{m,(6.77)} e^{-\delta_1 T/10}\|W\|_{L^2_{m,\delta}} \tag{6.77}
$$

for $\ell>0$ and $0\le \ell,n\le m-2$. (Here we assume W is T,ρ independent.) Note (6.77) holds at $\rho\in V(\rho'_0)$. Covering the compact set $\overline{V_1(\epsilon_2)\times_L V_2(\epsilon_2)}$ by finitely many such open sets $V(\rho_0')$, the inequality (6.77) holds at any $\rho \in V_1(\epsilon_2) \times_L V_2(\epsilon_2)$, by increasing the constant $C_{m,(6.77)}$ if necessary.

By definition we have

$$\begin{aligned}&(V^{\rho}_{T,1,(\kappa+1)}, V^{\rho}_{T,2,(\kappa+1)}, \Delta p^{\rho}_{T,(\kappa+1)})\\&= (I^0_{(\kappa),\rho,T} \circ \overline{D}^{-1}_{(\kappa),\rho,T} \circ (I^1_{(\kappa),\rho,T})^{-1})(\mathrm{Err}^{\rho}_{1,T,(\kappa)}, \mathrm{Err}^{\rho}_{2,T,(\kappa)}).\end{aligned}$$

We remark that $(I^1_{(\kappa),\rho,T})^{-1}(\mathrm{Err}^{\rho}_{1,T,(\kappa)}, \mathrm{Err}^{\rho}_{2,T,(\kappa)})$ is estimated by (6.10) since the isomorphism (6.16) is nothing but $(I^1_{i,(\kappa),\rho,T})^{-1}$.

Since $\Psi_{i;(\kappa-1)}(\rho, T)$ in (6.14) is nothing but $(I^0_{i,(\kappa),\rho,T})^{-1}$, the estimate (6.9) is one for $(I^0_{(\kappa),\rho,T})^{-1}(V^{\rho}_{T,1,(\kappa+1)}, V^{\rho}_{T,2,(\kappa+1)}, \Delta p^{\rho}_{T,(\kappa+1)})$.

Therefore, using Lemma 6.22, we derive (6.9) for $\kappa + 1$, by choosing $T_{2,m,\epsilon(6)}$ such that $C_{m,(6.77)} e^{-\delta_1 T_{2,m,\epsilon(6)}/10} \le C_{5,m}\mu$.

The proof of Proposition 6.9 (2) is complete. Thus the proof of Theorem 6.4 is now complete. $\square$

CHAPTER 7

Surjectivity and injectivity of the gluing map

In this chapter we prove surjectivity and injectivity of the map Glue_T in Theorem 3.13 and complete the proof of Theorem 3.13.[1] The proof goes along the line of [**D1**]. (See also [**FU**].) The surjectivity proof along the line of this chapter is written in [**FOn**, Section 14] and injectivity is proved in the same way. ([**FOn**, Section 14] studies the case of pseudoholomorphic curve without boundary. It however can be easily adapted to the bordered case as we mentioned in [**FOOO1**, page 417 lines 21-26].) Here we explain the argument in our situation in more detail.

We begin with the following a priori estimate.

PROPOSITION 7.1. *There exist $\epsilon_5, C_{m,(7.1)}$ such that if $u : (\Sigma_T, \partial\Sigma_T) \to (X, L)$ is an element of $\mathcal{M}^{\mathcal{E}_1\oplus\mathcal{E}_2}((\Sigma_T, \vec{z}); \beta)_\epsilon$ for $0 < \epsilon < \epsilon_5$ then we have*

$$\left\| \frac{\partial u}{\partial \tau} \right\|_{C^m([\tau-1,\tau+1]\times[0,1])} \le C_{m,(7.1)} e^{-\delta_1(5T-|\tau|)}. \tag{7.1}$$

PROOF. We first recall that elements in $\mathcal{E}_i$ are supported in K_i respectively. Therefore $u \in \mathcal{M}^{\mathcal{E}_1\oplus\mathcal{E}_2}((\Sigma_T, \vec{z}); \beta)_\epsilon$ satisifies $\overline{\partial} u = 0$ on the neck region $[-5T, 5T] \times [0, 1]$. Then (7.1) follows from Lemma 2.5. $\square$

We also have the following:

LEMMA 7.2. *$\mathcal{M}^{\mathcal{E}_1\oplus\mathcal{E}_2}((\Sigma_T, \vec{z}); \beta)_\epsilon$ is a smooth manifold of dimension* $\dim V_1 + \dim V_2 - \dim L$.

PROOF. Assumption 3.12 and the Mayer-Vietoris principle ([**Mr**]) imply that the linearized operator:

$$W^2_{m+1,\delta}((\Sigma_T, \partial\Sigma_T); u^*TX, u^*TL) \to L^2_{m,\delta}(\Sigma_T; u^*TX \otimes \Lambda^{0,1})/(\mathcal{E}_1(u) \oplus \mathcal{E}_2(u))$$

of the equation (3.19), which is defined by

$$V \mapsto (D_u\overline{\partial})(V) - (D_u\mathcal{E}_1)(\mathfrak{e}_1, V) - (D_u\mathcal{E}_2)(\mathfrak{e}_2, V) \mod \mathcal{E}_1(u) \oplus \mathcal{E}_2(u)$$

is surjective. (See [**FOOO1**, Proposition 7.1.27].) Here $\overline{\partial} u = (\mathfrak{e}_1, \mathfrak{e}_2) \in \mathcal{E}_1(u) \oplus \mathcal{E}_2(u)$. The lemma then follows from the implicit function theorem to solve the equation $\overline{\partial} u' \in \mathcal{E}_1(u) \oplus \mathcal{E}_2(u)$, for u'. $\square$

PROOF OF SURJECTIVITY. Let $u : (\Sigma_T, \partial\Sigma_T) \to (X, L)$ be an element of the moduli space $\mathcal{M}^{\mathcal{E}_1\oplus\mathcal{E}_2}((\Sigma_T, \vec{z}); u_1, u_2)_\epsilon$. The purpose here is to show that u lies in

[1] Here surjectivity means the second half of the statement of Theorem 3.13, that is 'The image contains $\mathcal{M}^{\mathcal{E}_1\oplus\mathcal{E}_2}((\Sigma_T, \vec{z}); u_1, u_2)_{\epsilon_{\epsilon(1),\epsilon(2),\langle\langle 3.13\rangle\rangle}}$.'

the image of Glue_T. For this purpose, we first decompose u into two pieces by defining two maps $u'_i : (\Sigma_i, \partial\Sigma_i) \to (X, L)$ as follows. We put $p_0^u = u(0,0) \in L$.

$$
\begin{aligned}
&u'_1(z) \\
&= \begin{cases} \mathrm{Exp}\,(p_0^u, \chi_{\mathcal{B}}^{\leftarrow}(\tau - T, t)\mathrm{E}(p_0^u, u(\tau,t))) & \text{if } z = (\tau,t) \in [-5T, 5T] \times [0,1] \\ u(z) & \text{if } z \in K_1 \\ p_0^u & \text{if } z \in [5T, \infty) \times [0,1]. \end{cases} \\
&u'_2(z) \\
&= \begin{cases} \mathrm{Exp}\,(p_0^u, \chi_{\mathcal{A}}^{\rightarrow}(\tau + T, t)\mathrm{E}(p_0^u, u(\tau,t))) & \text{if } z = (\tau,t) \in [-5T, 5T] \times [0,1] \\ u(z) & \text{if } z \in K_2 \\ p_0^u & \text{if } z \in (-\infty, -5T] \times [0,1]. \end{cases}
\end{aligned}
\tag{7.2}
$$

Proposition 7.1 implies

$$
\|\Pi^{\perp}_{\mathcal{E}_i(u'_i)} \overline{\partial} u'_i\|_{L^2_{m,\delta}(\Sigma_i)} \le C_{m,(7.3)} e^{-\delta_1 T}. \tag{7.3}
$$

On the other hand, by assumption and elliptic regularity we have

$$
\|u'_i - u_i\|_{W^2_{m+1,\delta}(\Sigma_i)} \le C_{m,(7.4)} \epsilon. \tag{7.4}
$$

Here we abuse the notation $u'_i - u_i = \mathrm{E}(u_i, u'_i)$ as we mentioned in Remark 2.2.

LEMMA 7.3. *There exist $\epsilon_{m,(7.5)}$ and $C_{m,(7.5)}$ with the following properties: If $u \in \mathcal{M}^{\mathcal{E}_1 \oplus \mathcal{E}_2}((\Sigma_T, \vec{z}); u_1, u_2)_\epsilon$ with $\epsilon < \epsilon_{m,(7.5)}$ then there exist $\rho_i \in V_i$ $(i = 1, 2)$ such that*

$$
\|\mathrm{E}(u_i^{\rho_i}, u'_i)\|_{W^2_{m+1,\delta}(\Sigma_i)} \le C_{m,(7.5)} e^{-\delta_1 T}, \tag{7.5}
$$

where $u_i^{\rho_i}$ is the map associated to ρ_i in $W^2_{m+1,\delta}((\Sigma_i, \partial\Sigma_i), (X, L))$.

PROOF. We consider the set of triples (v_1, v_2, q) where $v_i : (\Sigma_i, \partial\Sigma_i) \to (X, L)$ are maps of $L^2_{m+1,loc}$ class and $q \in L$ such that the section $\mathrm{E}(q, v_i)$ of T_qX has finite $L^2_{m+1,\delta}$ norm on $[0, \infty)_{\tau'} \times [0,1]$ $(i = 1)$, $(-\infty, 0]_{\tau''} \times [0,1]$ $(i = 2)$. The space of such (v_1, v_2, q) becomes a Hilbert manifold, which we denote by $\mathscr{L}$.

Suppose $d(v_i, u_i) \le \iota'_X$ pointwise. We define

$$
G_i(v_1, v_2, q) = \mathcal{P}(\overline{\partial} v_i) \in L^2_{m,\delta}(\Sigma_i, u_i^* TX \otimes \Lambda^{0,1}),
$$

where

$$
\mathcal{P} : L^2_{m,\delta}(\Sigma_i, v_i^* TX \otimes \Lambda^{0,1}) \to L^2_{m,\delta}(\Sigma_i, u_i^* TX \otimes \Lambda^{0,1})
$$

is defined by using parallel transport along the minimal geodesic. Then

$$
(G_1, G_2) : \mathscr{U} \to \bigoplus_{i=1}^{2} L^2_{m,\delta}(\Sigma_i, u_i^* TX \otimes \Lambda^{0,1})
$$

is a smooth map. Here $\mathscr{U}$ is a small neighborhood of (u_1, u_2, p_0) in $\mathscr{L}$.

We also define a smooth map

$$
\mathscr{E}_i : \mathscr{U} \times \mathcal{E}_i^{\mathfrak{ob}} \to L^2_{m,\delta}(\Sigma_i, u_i^* TX \otimes \Lambda^{0,1})
$$

by composing the two maps induced by the parallel transports

$$
\begin{aligned}
&L^2_{m,\delta}(\Sigma_i, (u_i^{\mathfrak{ob}})^* TX \otimes \Lambda^{0,1}) \to L^2_{m,\delta}(\Sigma_i, v_i^* TX \otimes \Lambda^{0,1}), \\
&L^2_{m,\delta}(\Sigma_i, v_i^* TX \otimes \Lambda^{0,1}) \to L^2_{m,\delta}(\Sigma_i, u_i^* TX \otimes \Lambda^{0,1}).
\end{aligned}
$$

(The map $\mathscr{E}_i$ is linear in the second factor.) By Assumption 3.12

$$(G_1, G_2) + (\mathscr{E}_1, \mathscr{E}_2) : \mathscr{U} \times \mathcal{E}_1^{\mathfrak{ob}} \times \mathcal{E}_2^{\mathfrak{ob}} \to \bigoplus_{i=1}^{2} L^2_{m,\delta}(\Sigma_i, u_i^* TX \otimes \Lambda^{0,1})$$

is transversal to zero. Moreover restriction of the projection $\mathscr{U} \times \mathcal{E}_1^{\mathfrak{ob}} \times \mathcal{E}_2^{\mathfrak{ob}} \to \mathscr{U}$ induces an isomorphism of

$$((G_1, G_2) + (\mathscr{E}_1, \mathscr{E}_2))^{-1}(0)$$

onto an open neighborhood of the origin in $V_1 \times_L V_2$. (Here the origin is the point corresponding to (u_1, u_2, p_0).) Finally we note that (7.4) implies $(u_1', u_2', p_0^u) \in \mathscr{U}$ and (7.3) implies

$$\|(\Pi^{\mathcal{E}_1^\perp \oplus \mathcal{E}_2^\perp})(G(u_1', u_2', p_0^u))\|_{L^2_{m,\delta}} \le Ce^{-\delta_1 T}.$$

Therefore there exists $\mathfrak{e}_i' \in \mathcal{E}_i^{\mathfrak{ob}}$ such that

$$\|(G_1, G_2)(u_1', u_2', p_0^u) + (\mathscr{E}_1, \mathscr{E}_2)(\mathfrak{e}_1', \mathfrak{e}_2')\| \le Ce^{-\delta_1 T}.$$

Now the lemma immediately follows from the implicit function theorem applied to the equation

$$(G_1, G_2)(v_1, v_2, q) + (\mathscr{E}_1, \mathscr{E}_2)(\mathfrak{e}_1, \mathfrak{e}_2) = 0.$$

□

We put $\rho = (\rho_1, \rho_2) \in V_1 \times_L V_2$. Then $u \in \mathcal{M}^{\mathcal{E}_1 \oplus \mathcal{E}_2}((\Sigma_T, \vec{z}); u_1, u_2)_\epsilon$ also implies

$$|\rho_i| \le C_{m,(7.6)}\epsilon. \tag{7.6}$$

By (7.5) we have

$$\|\mathrm{E}(u_T^\rho, u)\|_{W^2_{m+1,\delta}(\Sigma_T)} \le C_{m,(7.7)}e^{-\delta_1 T}. \tag{7.7}$$

Here $u_T^\rho = \mathrm{Glue}_T(\rho)$.

We put $V = \mathrm{E}(u_T^\rho, u) \in \Gamma((\Sigma_T, \partial\Sigma_T); (u_T^\rho)^*TX; (u_T^\rho)^*TL)$. Then

$$u(z) = \mathrm{Exp}(u_T^\rho(z), V(z)).$$

For $s \in [0,1]$ we define $u^s : (\Sigma_T, \partial\Sigma_T) \to (X, L)$ by

$$u^s(z) = \mathrm{Exp}(u_T^\rho(z), sV(z)). \tag{7.8}$$

(7.7) implies

$$\|\Pi^\perp_{(\mathcal{E}_1 \oplus \mathcal{E}_2)(u^s)} \overline{\partial} u^s\|_{L^2_{m,\delta}(\Sigma_T)} \le C_{m,(7.9)}e^{-\delta_1 T} \tag{7.9}$$

and

$$\left\|\frac{\partial}{\partial s}u^s\right\|_{W^2_{m+1,\delta}(K_i^S)} \le C_{m,(7.10)}e^{-\delta_1 T} \tag{7.10}$$

for each $s \in [0,1]$.

Lemma 7.4. *If T is sufficiently large, then there exists $\tilde{u}^s : (\Sigma_T, \partial\Sigma_T) \to (X, L)$ $(s \in [0,1])$ with the following properties.*

(1)

$$\overline{\partial}\tilde{u}^s \equiv 0 \mod (\mathcal{E}_1 \oplus \mathcal{E}_2)(\tilde{u}^s).$$

(2)

$$\left\|\frac{\partial}{\partial s}\tilde{u}^s\right\|_{W^2_{m+1,\delta}(K_i^S)} \le C_{m,S,(7.11)}e^{-\delta_1 T}. \tag{7.11}$$

(3) $\tilde{u}^s = u^s$ *for* $s = 0, 1$.

PROOF. Run the alternating method described in Chapter 5 in the one-parameter family version. Since u^s is already a solution for $s = 0, 1$, the Newton's iteration in Chapter 5 does not change it. More precisely, we regard u^s in (7.8) as $u^{\rho}_{T,(0)} = u^s_{T,(0)}$ and start our inductive construction at Step 0-3 (Lemma 5.2).

Then for $s = 0, 1$,

$$\mathrm{Err}^s_{1,T,(0)} = \chi^{\leftarrow}_{\mathcal{X}}(\overline{\partial} u^s_{T,(0)} - \mathfrak{e}^s_{1,T,(0)}) = 0.$$

We can show $\mathrm{Err}^s_{2,T,(0)} = 0$ in the same way. (Here $\mathrm{Err}^s_{i,T,(0)}$ is as in Definition 5.5.) Hence $V^s_{T,1,(1)} = V^s_{T,2,(1)} = 0$, $\Delta p^{\rho}_{T,(1)} = 0$ for $s = 0, 1$ by (5.30). Therefore $u^s_{T,(1)} = u^s_{T,(0)}$ by Definition 5.16. Thus $u^s_{T,(\kappa)} = u^s_{T,(0)} = u^s$ for all κ and $s = 0, 1$.

Then $\tilde{u}^s = \lim_{\kappa} u^s_{T,(\kappa)} = u^s$ for $s = 0, 1$ follows. □

LEMMA 7.5. *The map* $\mathrm{Glue}_T : V_1(\epsilon) \times_L V_2(\epsilon) \to \mathcal{M}^{\mathcal{E}_1 \oplus \mathcal{E}_2}((\Sigma_T, \vec{z}); u_1, u_2)_{\epsilon}$ *is an immersion if* $\epsilon < \epsilon_2$ *and* T *is sufficiently large.*

PROOF. We consider the composition of Glue_T with

$$\mathcal{M}^{\mathcal{E}_1 \oplus \mathcal{E}_2}((\Sigma_T, \vec{z}); u_1, u_2)_{\epsilon} \to L^2_{m+1}((K^S_i, K^S_i \cap \partial \Sigma_i), (X, L))$$

defined by restriction. In the case $T = \infty$ this composition is obtained by restriction of maps. By the unique continuation, this is certainly an immersion for $T = \infty$. Then the exponential decay property stated in Theorem 6.4 implies that it is an immersion for sufficiently large T. □

We are now ready to prove u lies in the image of Glue_T by showing that the following set

$$A = \{s \in [0, 1] \mid \tilde{u}^r \in \text{image of } \mathrm{Glue}_T \text{ for } r \le s\}$$

is nonempty, open and closed in $[0, 1]$. Obviously $0 \in A$ since $\widetilde{u}^0 = u^{\rho}_T$ by Lemma 7.4 (3). Lemma 7.2 implies that $\mathcal{M}^{\mathcal{E}_1 \oplus \mathcal{E}_2}((\Sigma_T, \vec{z}); u_1, u_2)_{\epsilon}$ is a smooth manifold and has the same dimension as $V_1 \times_L V_2$. Therefore Lemma 7.5 implies that A is open. The closedness of A follows from Lemma 7.5. In fact we may shrink the domain of Glue_T a bit if necessary so that it is not only an immersion but also the first derivative of its local inverse is uniformly bounded. Then if s is in the closure of A, the path $r \mapsto u^{\rho}_r$ ($r \in [0, s)$) has a lift to the domain of Glue_T with length $\le Ce^{-\delta_1 T}$. (We use Lemma 7.4 (2) here.) Therefore the closure of the lift is contained in the domain of Glue_T, when we take the constant $\epsilon_{\epsilon(1),\epsilon(2),\langle\langle 3.13\rangle\rangle}$ in Theorem 3.13 (2) sufficiently small. It implies $s \in A$.

Therefore $1 \in A$. Namely u lies in the image of Glue_T as required. □

PROOF OF INJECTIVITY. Let $\rho^j = (\rho^j_1, \rho^j_2) \in V_1 \times_L V_2$ for $j = 0, 1$. We assume

$$\mathrm{Glue}_T(\rho^0) = \mathrm{Glue}_T(\rho^1) \tag{7.12}$$

and

$$\|\rho^j_i\| < \epsilon. \tag{7.13}$$

We will prove that $\rho^0 = \rho^1$ if T is sufficiently large and ϵ is sufficiently small. We may assume that $V_1 \times_L V_2$ is connected and simply connected, by taking ϵ sufficiently small.

Then, we can take a path $s \mapsto \rho^s = (\rho^s_1, \rho^s_2) \in V_1 \times_L V_2$ such that

(1) $\rho^s = \rho^j$ for $s = j$, $j = 0, 1$.

(2)

$$\left\| \frac{\partial}{\partial s} \rho^s \right\| \le \Phi_1(\epsilon) + \Psi_1(T)$$

where $\lim_{\epsilon \to 0} \Phi_1(\epsilon) = 0$, $\lim_{T\to\infty} \Psi_1(T) = 0$.

We put $V(s) = \mathrm{E}(u_T^{\rho^0}, u_T^{\rho^s}) \in \Gamma((\Sigma_T, \partial\Sigma_T); (u_T^{\rho^0})^*TX; (u_T^{\rho^0})^*TL)$. Then

$$u_T^{\rho^s}(z) = \mathrm{Exp}(u_T^{\rho^0}(z), V(s)(z)).$$

(By (2) $u_T^{\rho^s}(z)$ is C^0-close to $u_T^{\rho^0}(z)$, as $\epsilon \to 0$. Therefore $V(s)$ is well defined if ϵ is small.) Note $V(1) = V(0)$ since $u^{\rho^1} = u^{\rho^0}$. Then for $w \in D^2 = \{w \in \mathbb{C} \mid |w| \le 1\}$, there exists $V(w)$ such that:

(1) $V(s) = V(w)$ if $w = e^{2\pi\sqrt{-1}s}$.

(2) We put $w = x + \sqrt{-1}y$.

$$\left\| \frac{\partial}{\partial x} V(w) \right\|_{W^2_{m+1,\delta}(\Sigma_T)} + \left\| \frac{\partial}{\partial y} V(w) \right\|_{W^2_{m+1,\delta}(\Sigma_T)} \le \Phi_2(\epsilon) + \Psi_2(T) \tag{7.14}$$

where $\lim_{\epsilon \to 0} \Phi_2(\epsilon) = 0$, $\lim_{T\to\infty} \Psi_2(T) = 0$.

We put $u^w(z) = \mathrm{Exp}(u_T^{\rho^0}(z), V(w)(z))$.

LEMMA 7.6. *If T is sufficiently large and ϵ is sufficiently small then there exists $\tilde{u}^w : (\Sigma_T, \partial\Sigma_T) \to (X, L)$ $(s \in [0,1])$ with the following properties.*

(1)

$$\overline{\partial} \tilde{u}^w \equiv 0 \mod (\mathcal{E}_1 \oplus \mathcal{E}_2)(\tilde{u}^w).$$

(2)

$$\left\| \frac{\partial}{\partial x} \tilde{u}^w \right\|_{W^2_{m+1,\delta}(K_i^S)} + \left\| \frac{\partial}{\partial y} \tilde{u}^w \right\|_{W^2_{m+1,\delta}(K_i^S)} \le \Phi_3(\epsilon) + \Psi_3(T) \tag{7.15}$$

with $\lim_{\epsilon \to 0} \Phi_3(\epsilon) = 0$, $\lim_{T\to\infty} \Psi_3(T) = 0$.

(3) $\tilde{u}^w = u^w$ *for* $w \in \partial D^2$.

PROOF. Run the alternating method described in Chapter 5 in the two-parameter family version. (3) is proved in the same way as Lemma 7.4. □

LEMMA 7.7. *If T is sufficiently large and ϵ is sufficiently small, there exists a smooth map $F : D^2 \to V_1 \times_L V_2$ such that*

(1) $\mathrm{Glue}_T(F(w)) = \tilde{u}^w$.

(2) *If $s \in [0,1]$ then we have:*

$$F(e^{2\pi\sqrt{-1}s}) = \rho^s.$$

PROOF. Note that $\rho \mapsto \mathrm{Glue}_T(\rho)$ is a local diffeomorphism. So we can apply the proof of homotopy lifting property as follows. Let $D^2(r) = \{z \in \mathbb{C} \mid |z - (r - 1)| \le r\}$. We put

$$A = \{r \in [0,1] \mid \exists\, F : D^2(r) \to V_1 \times_L V_2 \text{ satisfying (1) above and } F(-1) = \rho^{1/2}\}.$$

Since $\mathrm{Glue}_T(\rho)$ is a local diffeomorphism, A is open. We can use (7.15) to show closedness of A. Then, since $0 \in A$, it follows that $1 \in A$. The proof of Lemma 7.7 is complete. □

Lemma 7.7 implies $\rho^0 = F(e^{2\pi\sqrt{-1}\cdot 0}) = F(e^{2\pi\sqrt{-1}\cdot 1}) = \rho^1$. The proof of Theorem 3.13 is now complete. □

CHAPTER 8

Exponential decay estimate implies smoothness of coordinate change

In this chapter we demonstrate the way we use Theorems 3.13 and 6.4 to prove smoothness of coordinate change of the Kuranishi structure of the moduli space of bordered stable maps. Here we provide an argument in the case when we glue two source curves which are non-singular and stable. The proof for the general case is given in [**FOOO2**, Part IV]. (See especially its Section 21.)

8.1. Including deformation of source curve

We first generalize Theorems 3.13 and 6.4 and include deformations of complex structures of the source curves $(\Sigma_i, \vec{z}_i)$. We consider the situation of Theorem 3.13. Note we assumed that each $(\Sigma_i, \vec{z}_i)$ is stable. We also remark that there is no automorphism of $(\Sigma_i, \vec{z}_i)$. (See Remark 3.10.)

Let g_i be the genus of Σ_i. For simplicity of the notation we assume that the boundary of Σ_i is connected. Let $k_i + 1$ be the number of boundary marked points. (For the sake of simplicity of notations we consider the case when we have only boundary marked points. The case when we also have interior marked points can be studied in the same way.) We denote by

$$\mathcal{M}_{g_i;k_i+1} \tag{8.1}$$

the moduli space of bordered and marked Riemann surfaces with the same topological type as $(\Sigma_i, \vec{z}_i)$. Let $\mathcal{CM}_{g_i;k_i+1}$ be its compactification which consists of stable marked bordered curves. (See [**FOOO**, Subsection 2.1.2] for its definition.)

Let $\mathcal{V}'_i$ $(i = 1, 2)$ be a neighborhood of $[\Sigma_i, \vec{z}_i]$ in $\mathcal{M}_{g_i;k_i+1}$ and

$$\pi_i : \mathcal{C}'_i \to \mathcal{V}'_i, \qquad \mathfrak{z}_j : \mathcal{V}'_i \to \mathcal{C}'_i, \quad (j = 0, \dots, k_i) \tag{8.2}$$

be the universal family. Namely $\mathcal{C}'_i$ has a fiberwise complex structure such that $(\pi_i^{-1}(\sigma), (\mathfrak{z}_0(\sigma), \dots, \mathfrak{z}_{k_i}(\sigma)))$ is a representative of $\sigma \in \mathcal{V}'_i$.

We may choose $\mathcal{V}'_i$ so small that the bundle (8.2) is topologically trivial. We fix a trivialization $\mathcal{C}'_i \cong |\Sigma_i| \times \mathcal{V}'_i$ such that π_i is the projection to the second factor and that

$$\mathfrak{z}_j : \mathcal{V}'_i \to \mathcal{C}'_i \to |\Sigma_i| \times \mathcal{V}'_i \to |\Sigma_i|$$

are constant maps. (Here $|\Sigma_i|$ is the bordered curve Σ_i with its complex structure forgotten. Hereafter we write Σ_i in place of $|\Sigma_i|$ by a slight abuse of notation.) We denote

$$\Sigma_i(\sigma) = \pi_i^{-1}(\sigma), \quad \vec{z}_i(\sigma) = (\mathfrak{z}_0(\sigma), \dots, \mathfrak{z}_{k_i}(\sigma)).$$

We will next define $\Sigma_1(\sigma_1) \#_T \Sigma_2(\sigma_2) = \Sigma_T^{\sigma}$, which is obtained from $\Sigma_1(\sigma_1)$ and $\Sigma_2(\sigma_2)$ by gluing. To specify the way to glue we need to fix families of coordinates at the 0-th marked points $z_{i,0} \in \Sigma_i$.

For our purpose it is useful to take an analytic family of coordinates, which we define below. (Definition 8.5.) To define it we start with the analogue for a closed Riemann surface. Let $\mathcal{M}^{\rm cl}_{g,\ell+1}$ be the moduli space of closed Riemann surface of genus g with $\ell+1$ marked points and $\mathcal{CM}^{\rm cl}_{g,\ell+1}$ its compactification consisting of stable curves. (See [**DM**].) Let

$$\pi : \mathcal{C}^{\rm cl}_{g,\ell+1} \to \mathcal{CM}^{\rm cl}_{g,\ell+1} \tag{8.3}$$

be the universal family. Let $[\Sigma, \vec z] \in \mathcal{CM}^{\rm cl}_{g,\ell+1}$ and Γ be the automorphism group of $[\Sigma, \vec z]$. Then a neighborhood of $[\Sigma, \vec z]$ in $\mathcal{CM}^{\rm cl}_{g,\ell+1}$ may be regarded as $\mathcal{V}/\Gamma$, where $\mathcal{V}$ is a complex manifold. $\pi^{-1}(\mathcal{V}/\Gamma)$ is identified with $\mathcal{C}^{\rm cl}(\mathcal{V})/\Gamma$ where $\mathcal{C}^{\rm cl}(\mathcal{V})$ is a complex manifold and $\pi_{\mathcal{V}} : \mathcal{C}^{\rm cl}(\mathcal{V}) \to \mathcal{V}$ is a Γ equivariant holomorphic map. It comes with Γ equivariant sections $\mathfrak{z}_j : \mathcal{V} \to \mathcal{C}^{\rm cl}(\mathcal{V})$, $j = 0, 1, \dots, \ell$ such that

$$(\Sigma(\sigma), \vec z(\sigma)) = (\pi^{-1}(\sigma), (\mathfrak{z}_0(\sigma), \dots, \mathfrak{z}_\ell(\sigma)))$$

is a representative of σ. We put $z_j(\sigma) = \mathfrak{z}_j(\sigma)$.

DEFINITION 8.1. A *complex analytic family of coordinates* at the j-th marked point on $\mathcal{V}$ is a map $\widetilde\varphi : D^2 \times \mathcal{V} \to \mathcal{C}^{\rm cl}(\mathcal{V})$ with the following properties.

(1) $\widetilde\varphi$ is a biholomorphic map onto its image, which is an open subset.
(2) $\pi \circ \widetilde\varphi : D^2 \times \mathcal{V} \to \mathcal{V}$ coincides with the projection to the second factor.
(3) $\widetilde\varphi(0, \sigma) = z_j(\sigma)$.
(4) We consider the Γ action on $T_{z_j}\Sigma \cong \mathbb{C}$ and use it to define a Γ action on D^2. Then $\widetilde\varphi$ is Γ invariant.

Definition 8.1 implies that the restriction of $\tilde\varphi$ to $D^2 \times \{\sigma\}$ becomes a complex coordinate of $\Sigma(\sigma)$ at $\mathfrak{z}_j(\sigma)$.

Existence of a complex analytic family of coordinates is a consequence of the Γ-equivariant version of implicit function theorem in complex analytic category and is well-known. (See Appendix H.)

Let $\mathcal{V}_i$ be neighborhoods of $[\Sigma^{\rm cl}_i, \vec z_i] \in \mathcal{M}^{\rm cl}_{g_i,\ell_i+1}$, for $i = 1, 2$. We assume that $[\Sigma^{\rm cl}_1, \vec z_1] \ne [\Sigma^{\rm cl}_2, \vec z_2]$.[1] We put $\Gamma_i = {\rm Aut}(\Sigma^{\rm cl}_i, \vec z_i)$. Let $\widetilde\varphi_i : D^2_i \times \mathcal{V}_i \to \mathcal{C}^{\rm cl}_i(\mathcal{V}_i)$ be two complex analytic families of coordinates at the 0-th marked points on neighborhoods $\mathcal{V}_i$ of $[\Sigma^{\rm cl}_i, \vec z_i]$.

We identify $z_{1,0} \in \Sigma^{\rm cl}_1$ with $z_{2,0} \in \Sigma^{\rm cl}_2$. Then we obtain an element

$$[\Sigma^{\rm cl}_\infty, \vec z_\infty] \in \mathcal{CM}^{\rm cl}_{g_1+g_2,\ell_1+\ell_2}.$$

Using our complex analytic families of coordinates we define a map

$$\text{Gluesoc} : \mathcal{V}_1 \times \mathcal{V}_2 \times D^2(\epsilon) \to \mathcal{CM}^{\rm cl}_{g_1+g_2,\ell_1+\ell_2} \tag{8.4}$$

as follows. (Here Gluesoc stands for "gluing source".)

Let $\sigma_i \in \mathcal{V}_i$ and $(\Sigma_i(\sigma_i), \vec z_i(\sigma_i))$ be a marked Riemann surface representing it. We define a map $\varphi_{i,\sigma_i} : D^2 \to \Sigma_i(\sigma_i)$ by

$$\varphi_{i,\sigma_i}(z) = \widetilde\varphi_i(z, \sigma_i). \tag{8.5}$$

For each given $\mathfrak{r} \in D^2(\epsilon)$. We consider the disjoint union

$$(\Sigma_1(\sigma_1) \setminus \varphi_{1,\sigma_1}(D^2(|\mathfrak{r}|))) \sqcup (\Sigma_2(\sigma_2) \setminus \varphi_{2,\sigma_2}(D^2(|\mathfrak{r}|))).$$

[1] In case $[\Sigma^{\rm cl}_1, \vec z_1] = [\Sigma^{\rm cl}_2, \vec z_2]$ we may have extra $\mathbb{Z}_2$ symmetry. Other than that the argument is the same as in the case $[\Sigma^{\rm cl}_1, \vec z_1] \ne [\Sigma^{\rm cl}_2, \vec z_2]$.

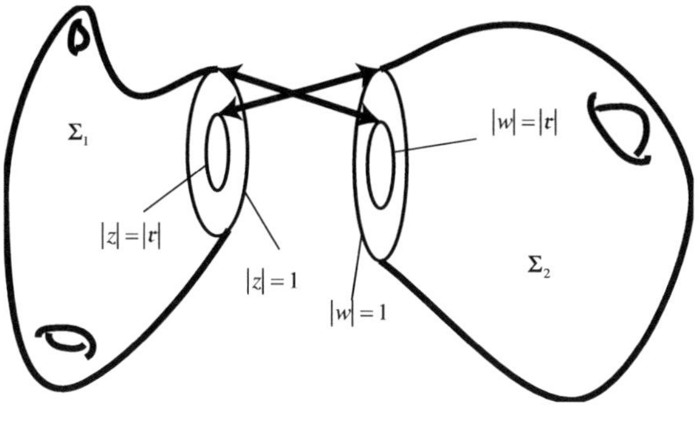

FIGURE 8.1. Source curve gluing.

(Here and hereafter $D^2(r) = \{z \in \mathbb{C} \mid |z| < r\}$.) We identify $\varphi_{1,\sigma_1}(z) \in \Sigma_1(\sigma_1) \setminus \varphi_{1,\sigma_1}(D^2(|\mathfrak{r}|))$ with $\varphi_{2,\sigma_2}(w) \in \Sigma_2(\sigma_2) \setminus \varphi_{2,\sigma_2}(D^2(|\mathfrak{r}|))$ if

$$zw = \mathfrak{r}.$$

See Figure 8.1.

By this identification we obtain a Riemann surface, which we denote by the symbol $\Sigma_1(\sigma_1)\#_{\mathfrak{r}}\Sigma_2(\sigma_2)$. We put

$$\text{Gluesoc}(\sigma_1, \sigma_2, \mathfrak{r}) = (\Sigma_1(\sigma_1)\#_{\mathfrak{r}}\Sigma_2(\sigma_2), \vec{z}_1'(\sigma_1) \cup \vec{z}_2'(\sigma_2)), \tag{8.6}$$

where $\vec{z}_i'(\sigma_i) = \vec{z}_i(\sigma_i) \setminus \{z_{i,0}(\sigma_i)\}$.

We define the Γ_i action on $D^2(\epsilon)$ by identifying $D^2(\epsilon)$ with the ball of radius ϵ centered at origin in the tangent space $T_{z_{i,0}}\Sigma_i$. Then Gluesoc is $\Gamma_1 \times \Gamma_2$ equivariant.

LEMMA 8.2. *The map* Gluesoc *induces a biholomorphic map*

$$\frac{\mathcal{V}_1 \times \mathcal{V}_2 \times D^2(\epsilon)}{\Gamma_1 \times \Gamma_2} \to \mathcal{CM}^{\text{cl}}_{g_1+g_2,\ell_1+\ell_2}$$

onto an open neighborhood of $[\Sigma^{\text{cl}}, \vec{z}]$.

PROOF. We define

$$\Omega_1 = \mathcal{V}_2 \times \coprod_{\mathfrak{r}\in D^2(\epsilon),\sigma_1\in\mathcal{V}_1} (\Sigma_1(\sigma_1) \setminus \varphi_{1,\sigma_1}(D^2(|\mathfrak{r}|))) \times \{\mathfrak{r}\}, \tag{8.7}$$

$$\Omega_2 = \mathcal{V}_1 \times \coprod_{\mathfrak{r}\in D^2(\epsilon),\sigma_2\in\mathcal{V}_2} (\Sigma_2(\sigma_2) \setminus \varphi_{2,\sigma_2}(D^2(|\mathfrak{r}|))) \times \{\mathfrak{r}\}. \tag{8.8}$$

We regard Ω_1 (resp. Ω_2) as an open subset of $\mathcal{V}_2 \times \mathcal{C}^{\text{cl}}(\mathcal{V}_1)$ (resp. $\mathcal{V}_1 \times \mathcal{C}^{\text{cl}}(\mathcal{V}_2)$). So they are complex manifolds. We also put

$$\Omega_3 = \{(z, w) \mid |z|, |w| < 1, \ |zw| < \epsilon\} \times \mathcal{V}_1 \times \mathcal{V}_2. \tag{8.9}$$

We identify $(z, w, \sigma_1, \sigma_2) \in \Omega_3$ with $(\sigma_2, \varphi_1^\sigma(z), zw) \in \Omega_1$ and with $(\sigma_1, \varphi_2^\sigma(w), zw) \in \Omega_2$. We obtain a complex manifold by this identification, which we denote by $\mathcal{C}(\mathcal{V}_1, \mathcal{V}_2, \epsilon)$. We define $\pi : \mathcal{C}(\mathcal{V}_1, \mathcal{V}_2, \epsilon) \to \mathcal{V}_1 \times \mathcal{V}_2 \times D^2(\epsilon)$ by

$$\begin{aligned} \pi(\sigma_2, \hat{z}, \mathfrak{r}) &= (\pi_1(\hat{z}), \sigma_2, \mathfrak{r}), && \text{on } \Omega_1, \\ \pi(\sigma_1, \hat{w}, \mathfrak{r}) &= (\sigma_1, \pi_2(\hat{w}), \mathfrak{r}), && \text{on } \Omega_2, \\ \pi((z, w), \sigma_1, \sigma_2) &= (\sigma_1, \sigma_2, zw), && \text{on } \Omega_3. \end{aligned} \tag{8.10}$$

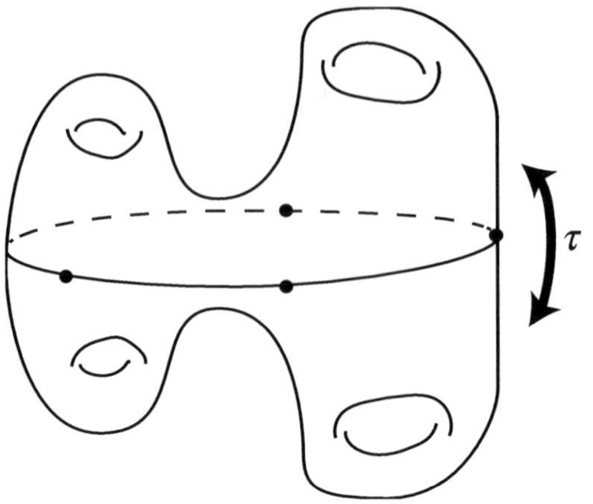

FIGURE 8.2. Doubling bordered curve.

Here $\hat{z} \in \mathcal{C}(\mathcal{V}_1)$, $\hat{w} \in \mathcal{C}(\mathcal{V}_2)$, $\mathfrak{r} \in D^2(\epsilon)$ and $\pi_i : \mathcal{C}(\mathcal{V}_i) \to \mathcal{V}_i$ is the projection. π is a well defined holomorphic map.

Sections $\mathfrak{z}_{i,j} : \mathcal{V}_i \to \mathcal{C}(\mathcal{V}_i)$, $j = 0, \dots, k_i$ (which give j-th marked point of the fiber) induce the sections $\mathfrak{z}_j : \mathcal{V}_1 \times \mathcal{V}_2 \times D^2(\epsilon) \to \mathcal{C}(\mathcal{V}_1, \mathcal{V}_2, \epsilon)$ for $j = 1, \dots, k_1 + k_2$ in an obvious way.

For $(\sigma_1, \sigma_2, \mathfrak{r}) \in \mathcal{V}_1 \times \mathcal{V}_2 \times D^2(\epsilon)$ it is easy to see from the definition that

$$(\pi^{-1}(\sigma_1, \sigma_2, \mathfrak{r}), (\mathfrak{z}_j(\sigma_1, \sigma_2, \mathfrak{r}))_{j=1,\dots,k_1+k_2})$$

is a representative of $\mathrm{Gluesoc}(\sigma_1, \sigma_2, \mathfrak{r})$.

We furthermore observe that $\pi : \mathcal{C}(\mathcal{V}_1, \mathcal{V}_2, \epsilon) \to \mathcal{V}_1 \times \mathcal{V}_2 \times D^2(\epsilon)$ is $\Gamma_1 \times \Gamma_2$ equivariant.

Therefore the biholomorphicity of the map Gluesoc is a consequence of the definition of the universal family. □

Now we go back to the case of bordered curves. Let $[\Sigma, \vec{z}] \in \mathcal{M}_{g;k+1}$. We take its *double* as in [**FOOO**, page 44] to obtain a closed Riemann surface, $[\Sigma^{\mathrm{cl}}, \vec{z}^{\mathrm{cl}}] \in \mathcal{M}^{\mathrm{cl}}_{2g;k+1}$. There exits an anti-holomorphic involution $\tau : \Sigma^{\mathrm{cl}} \to \Sigma^{\mathrm{cl}}$ such that :

(a) The fixed point set $\Sigma^{\mathrm{cl}}_{\mathbb{R}}$ of τ is S^1, which contains all the marked points.
(b) The complement $\Sigma^{\mathrm{cl}} \setminus \Sigma^{\mathrm{cl}}_{\mathbb{R}}$ consists of two connected components. The closure of one of them with marked points is biholomorphic to $(\Sigma, \vec{z})$.

See Figure 8.2. Let $\pi : \mathcal{C}(\mathcal{V}) \to \mathcal{V}$ be the universal family in a neighborhood of $[\Sigma^{\mathrm{cl}}, \vec{z}^{\mathrm{cl}}]$.

LEMMA 8.3. *There exist anti-holomorphic involutions $\tau : \mathcal{C}(\mathcal{V}) \to \mathcal{C}(\mathcal{V})$, $\tau : \mathcal{V} \to \mathcal{V}$ with the following properties.*

(1) $\pi \circ \tau = \tau \circ \pi$.
(2) *The real dimension of the fixed point set $\mathcal{V}^{\mathbb{R}}$ of $\tau : \mathcal{V} \to \mathcal{V}$ is equal to the complex dimension of $\mathcal{V}$. $\mathcal{V}^{\mathbb{R}}$ is identified with an open neighborhood of $[\Sigma, \vec{z}]$ in $\mathcal{M}_{g;k+1}$.*
(3) $\sigma_0 = [\Sigma^{\mathrm{cl}}, \vec{z}^{\mathrm{cl}}] \in \mathcal{V}$ *is a fixed point of* τ.
(4) *The restriction of $\tau : \mathcal{V} \to \mathcal{V}$ to the fiber of σ_0 coincides with the map $\tau : \Sigma^{\mathrm{cl}} \to \Sigma^{\mathrm{cl}}$.*

(5) *We restrict the universal family to the fixed point set* $\mathcal{V}^{\mathbb{R}}$ *of* $\tau : \mathcal{V} \to \mathcal{V}$. *We obtain a family of bordered marked Riemann surfaces in the same way as (b) applied to each of the fiber of* $\sigma \in \mathcal{V}^{\mathbb{R}}$. *This family is the universal family on* $\mathcal{V}^{\mathbb{R}} \subset \mathcal{M}_{g;k+1}$ *of bordered Riemann surfaces.*

PROOF. By [**DM**], $\mathcal{C}^{\rm cl} \to \mathcal{M}^{\rm cl}_{g;k+1}$ is a morphism in the category of the stacks defined over $\mathbb{R}$. The marked curve $[\Sigma^{\rm cl}, \vec z^{\rm cl}]$ together with $\tau : \Sigma^{\rm cl} \to \Sigma^{\rm cl}$ defines an $\mathbb{R}$-valued point of $\mathcal{M}^{\rm cl}_{g;k+1}$. The lemma is a consequence of this fact. □

LEMMA 8.4. *In the situation of Lemma* 8.3 *there exists a complex analytic family of coordinates at the* j*-th marked point,* $\varphi : D^2 \times \mathcal{V} \to C^{\rm cl}(\mathcal{V})$, *in the sense of Definition* 8.1 *such that*

$$\varphi(\overline{z}, \tau(\sigma)) = \tau(\varphi(z, \sigma)) \tag{8.11}$$

in addition.

PROOF. In place of $\Gamma = \mathrm{Aut}(\Sigma^{\rm cl}, \vec z^{\rm cl})$ we consider its $\mathbb{Z}_2$ extension $\Gamma_+ = \Gamma \cup \{\tau\gamma \mid \gamma \in \Gamma\}$. Then the proof is the same as the proof of existence of a complex analytic family of coordinates, which uses an equivariant version of the implicit function theorem. For completeness' sake, we provide the detail of the proof in Appendix H. □

We take a complex analytic family of coordinates $\varphi : D^2 \times \mathcal{V} \to C^{\rm cl}(\mathcal{V})$ as in Lemma 8.4. Note $\Sigma = \Sigma(\sigma_0)$ is the closure of one of the connected components of $\Sigma^{\rm cl} \setminus \Sigma^{\rm cl}_{\mathbb{R}}$. Replacing φ by the map $(z, \sigma) \mapsto \varphi(-z, \sigma)$ if necessary we may assume

$$\varphi(D^2_{\le 0}(1) \times \{\sigma_0\}) \subset \Sigma.$$

(Here and hereafter we write $D^2_{\le 0}(r) = \{z \in D^2(r) \mid \mathrm{Im} z \le 0\}$.) Then for any $\sigma \in \mathcal{V}^{\mathbb{R}}$ we have

$$\varphi(D^2_{\le 0}(1) \times \{\sigma\}) \subset \Sigma(\sigma).$$

We identify the bordered Riemann surface $\Sigma(\sigma)$ as a connected component of $\Sigma^{\rm cl}(\sigma) \setminus \Sigma^{\rm cl}_{\mathbb{R}}(\sigma)$ by Lemma 8.3 (5).

We thus obtain

$$\varphi^{\mathbb{R}} : D^2_{\le 0}(1) \times \mathcal{V}^{\mathbb{R}} \to \mathcal{C}(\mathcal{V}^{\mathbb{R}}). \tag{8.12}$$

Here $\pi : \mathcal{C}(\mathcal{V}^{\mathbb{R}}) \to \mathcal{V}^{\mathbb{R}}$ is the universal family of bordered Riemann surfaces. For each $\sigma \in \mathcal{V}^{\mathbb{R}}$ the map $\varphi^{\mathbb{R}}$ determines a (complex) coordinate at the j-th marked point of the bordered curve $\Sigma(\sigma)$.

DEFINITION 8.5. We define an *analytic family of coordinates* at the j-th marked point on $\mathcal{V}^{\mathbb{R}}$ to be a map $\varphi^{\mathbb{R}}$ as in (8.12) obtained from the complex analytic family of coordinates satisfying the conclusion of Lemma 8.4.

Let $[\Sigma_i, \vec z_i] \in \mathcal{M}_{g_i;k_i+1}$ for $i = 1, 2$. We take their doubles $[\Sigma_i^{\rm cl}, \vec z_i^{\rm cl}] \in \mathcal{M}^{\rm cl}_{g_i;k_i+1}$. Let $\mathcal{C}(\mathcal{V}_i) \to \mathcal{V}_i$ be the universal family on a neighborhood $\mathcal{V}_i$ of $[\Sigma_i^{\rm cl}, \vec z_i^{\rm cl}]$ which satisfies the conclusion of Lemma 8.3. We take $\varphi_i : D^2(1) \times \mathcal{V}_i \to \mathcal{C}(\mathcal{V}_i)$ which is a complex analytic family of coordinates at the 0-th marked points. We assume that φ_i satisfies the conclusion (8.11) of Lemma 8.4. We then obtain analytic families of coordinates at 0-th marked points on $\mathcal{V}_i^{\mathbb{R}}$,

$$\varphi_i^{\mathbb{R}} : D^2_{\le 0}(1) \times \mathcal{V}_i^{\mathbb{R}} \to \mathcal{C}(\mathcal{V}_i^{\mathbb{R}}). \tag{8.13}$$

We define $\varphi^{\mathbb{R}}_{i,\sigma_i} : D^2_{\le 0} \to \Sigma_i(\sigma_i)$ by

$$\varphi^{\mathbb{R}}_{i,\sigma_i}(z) = \varphi^{\mathbb{R}}_i(z,\sigma_i). \tag{8.14}$$

Let $r \in [0,\epsilon)$, $(\sigma_1,\sigma_2) \in \mathcal{V}^{\mathbb{R}}_1 \times \mathcal{V}^{\mathbb{R}}_2$. We define $\Sigma_1(\sigma_1)\#_r\Sigma_2(\sigma_2)$ as follows. Let $z_{i,0}(\sigma_i) \in \Sigma_i(\sigma_i)$ be the 0-th marked point. We consider the disjoint union

$$\left(\Sigma_1(\sigma_1) \setminus \varphi^{\mathbb{R}}_{1,\sigma_1}(D^2_{\le 0}(r))\right) \sqcup \left(\Sigma_2(\sigma_2) \setminus \varphi^{\mathbb{R}}_{2,\sigma_2}(D^2_{\le 0}(r))\right).$$

We define an equivalence relation on this set such that

$$\varphi^{\mathbb{R}}_{1,\sigma_1}(z) \in \Sigma_1(\sigma_1) \setminus \varphi^{\mathbb{R}}_{1,\sigma_1}(D^2_{\le 0}(r)),$$

is equivalent to $\varphi^{\mathbb{R}}_{2,\sigma_2}(w) \in \Sigma_2(\sigma_2) \setminus \varphi^{\mathbb{R}}_{2,\sigma_2}(D^2_{\le 0}(r))$ if and only if

$$zw = -r. \tag{8.15}$$

Let $\Sigma_1(\sigma_1)\#_r\Sigma_2(\sigma_2)$ be the set of the equivalence classes of the above equivalence relation. It becomes a bordered curve. Thus we have defined

$$\mathrm{Gluesoc}^{\mathbb{R}} : \mathcal{V}^{\mathbb{R}}_1 \times \mathcal{V}^{\mathbb{R}}_2 \times [0,\epsilon) \to \mathcal{CM}_{g_1+g_2,\ell_1+\ell_2}. \tag{8.16}$$

The next diagram commutes:

$$\begin{array}{ccc} \mathcal{V}^{\mathbb{R}}_1 \times \mathcal{V}^{\mathbb{R}}_2 \times [0,\epsilon) & \xrightarrow{\mathrm{Gluesoc}^{\mathbb{R}}} & \mathcal{CM}_{g_1+g_2,\ell_1+\ell_2} \\ \downarrow & & \downarrow \\ \mathcal{V}_1 \times \mathcal{V}_2 \times D^2(\epsilon) & \xrightarrow{\mathrm{Gluesoc}} & \mathcal{CM}^{\mathrm{cl}}_{g_1+g_2,\ell_1+\ell_2} \end{array} \tag{8.17}$$

Here $[0,\epsilon) \to D^2(\epsilon)$ in the first vertical arrow is $r \mapsto -r$. The other parts of the vertical arrows are obvious inclusions.

The source curve gluing map $\mathrm{Gluesoc}^{\mathbb{R}}$ can be identified with the one we described at the beginning of Chapter 3 as follows. We put

$$\exp(-10\pi T) = r, \qquad K_i(\sigma_i) = \Sigma_i(\sigma) \setminus \varphi^{\mathbb{R}}_{i,\sigma_i}(D^2_{\le 0}). \tag{8.18}$$

We define

$$\Sigma_1(\sigma_1) \setminus K_1(\sigma_1) \cong [0,\infty)_{\tau'} \times [0,1], \quad \Sigma_2(\sigma_2) \setminus K_2(\sigma_2) \cong (-\infty,0]_{\tau''} \times [0,1],$$

by

$$\varphi^{\mathbb{R}}_{1,\sigma_1}(e^{\pi(x+\sqrt{-1}y)}) \mapsto (-x,-y) = (\tau',t), \ \varphi^{\mathbb{R}}_{2,\sigma_2}(e^{\pi(x+\sqrt{-1}(y+1))}) \mapsto (x,y) = (\tau'',t).$$

Then $\Sigma_1(\sigma_1)\#_T\Sigma_2(\sigma_2)$ as in (3.2) is isomorphic to $\Sigma_1(\sigma_1)\#_r\Sigma_2(\sigma_2)$.

In fact the variables z, w appearing in (8.15) are related to the coordinates τ', τ'', t of the neck region we used in Chapter 6 by

$$z = e^{-\pi(\tau'+\sqrt{-1}t)}, \quad w = -e^{\pi(\tau''+\sqrt{-1}t)}.$$

Note $zw = -r = -e^{-10\pi T}$ is equivalent to $\tau'' = \tau' - 10T$. See Figure 8.3,

DEFINITION 8.6. For $[\Sigma_i, \vec{z}_i]$ in $\mathcal{M}_{g_i;k_i+1}$ we consider the following set Ξ_i of data:

(1) A neighborhood $\mathcal{V}^{\mathbb{R}}_i$ of $[\Sigma_i, \vec{z}_i]$ in $\mathcal{M}_{g_i;k_i+1}$.
(2) An analytic family of coordinates $\varphi^{\mathbb{R}}_i$ at the 0-th marked points as in (8.13).

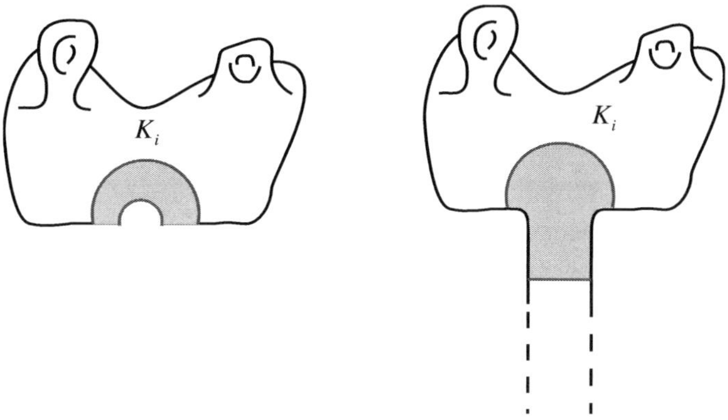

FIGURE 8.3. Coordinate at infinity of bordered curve.

(3) A trivialization of the universal family bundle $\mathcal{C}(\mathcal{V}_i^{\mathbb{R}}) \to \mathcal{V}_i^{\mathbb{R}}$ in the C^∞ category. It is a diffeomorphism $\mathcal{C}(\mathcal{V}_i^{\mathbb{R}}) \cong \Sigma_i \times \mathcal{V}_i^{\mathbb{R}}$ that commutes with the projections. Moreover we require it to commute with $\varphi_i^{\mathbb{R}}$. Namely the composition

$$D^2_{\le 0}(1) \times \mathcal{V}_i^{\mathbb{R}} \to \mathcal{C}(\mathcal{V}_i^{\mathbb{R}}) \to \Sigma_i \times \mathcal{V}_i^{\mathbb{R}}$$

of $\varphi_i^{\mathbb{R}}$ and the trivialization is of the form $(z, \sigma) \mapsto (\phi(z), \sigma)$ where $\phi : D^2_{\le 0}(1) \to \Sigma_i$ is independent of σ.

A *gluing datum* centered at $([\Sigma_1, \vec{z}_1], [\Sigma_2, \vec{z}_2])$ is by definition a pair $\Xi = (\Xi_1, \Xi_2)$ such that Ξ_i are given as above.

DEFINITION-LEMMA 8.7. Let $\Xi = (\Xi_1, \Xi_2)$ be a gluing datum centered at $([\Sigma_1, \vec{z}_1], [\Sigma_2, \vec{z}_2])$.

(1) It induces the map (8.16). We call the latter the *source gluing map associated to* Ξ and write $\mathrm{Gluesoc}^{\mathbb{R}}_{\Xi}$.
(2) For any $\sigma = (\sigma_1, \sigma_2) \in \mathcal{V}_1^{\mathbb{R}} \times \mathcal{V}_2^{\mathbb{R}}$, it induces a holomorphic embedding

$$\tilde{\mathfrak{I}}^{\sigma}_{\Xi,i} : K_i(\sigma_i) \to \Sigma_1(\sigma_1) \#_r \Sigma_2(\sigma_2).$$

We call it the *canonical holomorphic embedding associated to* Ξ. Here the bordered curve $\Sigma_1(\sigma_1) \#_r \Sigma_2(\sigma_2)$ together with marked points represents the element $\mathrm{Gluesoc}^{\mathbb{R}}_{\Xi}(\sigma_1, \sigma_2, r)$.
(3) For any $\sigma = (\sigma_1, \sigma_2) \in \mathcal{V}_1^{\mathbb{R}} \times \mathcal{V}_2^{\mathbb{R}}$, it also induces a smooth embedding

$$\mathfrak{I}^{\sigma}_{\Xi,i} : K_i \to \Sigma_1(\sigma_1) \#_r \Sigma_2(\sigma_2).$$

We call it the *canonical embedding associated to* Ξ. Recall $K_i \subset \Sigma_i$.

PROOF. We discussed (1) already. Statement (2) is a consequence of (8.18). Statement (3) follows from (2) and the trivialization of the universal family bundle given in Definition 8.6 (3). □

8.2. Obstruction bundle and family gluing map

In this section, we will formulate a family version of the gluing process of pseudoholomorphic maps over the moduli space of stable curves.

We first recall the definitions of stable maps and of their moduli space.

DEFINITION 8.8. Two marked pseudoholomorphic maps $((\Sigma,\vec{z}),u)$, $((\Sigma',\vec{z}'),u')$ are said to be isomorphic, if there exists an isomorphism $\varphi:(\Sigma,z)\to(\Sigma',\vec{z}')$ such that $u'=u\circ\varphi^{-1}$. A self-isomorphism $\varphi:(\Sigma,\vec{z})\to(\Sigma,\vec{z})$ is called an automorphism of $(\Sigma,\vec{z})$ if $u=u\circ\varphi^{-1}$. We denote

$$\mathrm{Aut}((\Sigma,\vec{z}),u)=\{\varphi\in\mathrm{Aut}(\Sigma,\vec{z})\mid u\circ\varphi=u\}.$$

We call the pair $((\Sigma,\vec{z}),u)$ a *stable map* if $\#\,\mathrm{Aut}((\Sigma,\vec{z}),u)$ is finite. We define the moduli space of the isomorphism classes of bordered stable maps $u:(\Sigma,\partial\Sigma)\to(X,L)$ of genus g with $k+1$ boundary marked points and homology class $\beta\in H_2(X,L;\mathbb{Z})$ and denote it by

$$\mathcal{M}_{g,k+1}(X,L;\beta).$$

See [**FOOO**, Definition 2.1.27].

We next define the obstruction spaces. Let $(\Sigma_i^{\mathrm{ob}},\vec{z}_i^{\mathrm{ob}})\in\mathcal{M}_{g_i,k_i+1}$, and let $u_i^{\mathrm{ob}}:(\Sigma_i^{\mathrm{ob}},\partial\Sigma_i^{\mathrm{ob}})\to(X,L)$ be a pseudoholomorphic map.[2]

DEFINITION 8.9. An *obstruction bundle datum* centered at

$$(((\Sigma_1^{\mathrm{ob}},\vec{z}_1^{\mathrm{ob}}),u_1^{\mathrm{ob}}),((\Sigma_2^{\mathrm{ob}},\vec{z}_2^{\mathrm{ob}}),u_2^{\mathrm{ob}}))$$

consists of the objects $(\Xi^{\mathrm{ob}},(\mathcal{E}_1^{\mathrm{ob}},\mathcal{E}_2^{\mathrm{ob}}))$ such that:

(1) $\Xi^{\mathrm{ob}}=(\Xi_1^{\mathrm{ob}},\Xi_2^{\mathrm{ob}})$ is a gluing datum centered at $((\Sigma_1^{\mathrm{ob}},\vec{z}_1^{\mathrm{ob}}),(\Sigma_2^{\mathrm{ob}},\vec{z}_2^{\mathrm{ob}}))$.
(2) $\mathcal{E}_i^{\mathrm{ob}}$ is a finite dimensional subspace of $\Gamma(\Sigma_i^{\mathrm{ob}};(u_i^{\mathrm{ob}})^*TX\otimes\Lambda^{0,1})$, the space of smooth sections. We assume that the supports of the elements of $\mathcal{E}_i^{\mathrm{ob}}$ are contained in $\mathrm{Int}\,K_i^{\mathrm{ob}}$. Here $K_i^{\mathrm{ob}}=\Sigma_i^{\mathrm{ob}}\setminus\varphi_{i,\mathrm{ob}}^{\mathbb{R}}(D^2_{\le 0}(1))$. Note the map $\varphi_{i,\mathrm{ob}}^{\mathbb{R}}$ is the analytic family of coordinates at the 0-th marked point which is a part of Ξ_i^{ob}.
(3) The maps u_1^{ob}, u_2^{ob} satisfy Assumption 3.12 . Here we replace Σ_i, $\vec{z}_i$, u_i, $\mathrm{ev}_{i,\infty}$ in Assumption 3.12 by Σ_i^{ob}, $\vec{z}_i^{\mathrm{ob}}$, u_i^{ob}, $\mathrm{ev}_{i,0}^{\mathrm{ob}}$.

We call $(((\Sigma_1^{\mathrm{ob}},\vec{z}_1^{\mathrm{ob}}),u_1^{\mathrm{ob}}),((\Sigma_2^{\mathrm{ob}},\vec{z}_2^{\mathrm{ob}}),u_2^{\mathrm{ob}}))$ the *obstruction center.*

We define the obstruction bundle $\mathcal{E}_i(u')$ for an element $u':(\Sigma',\partial\Sigma')\to(X,L)$ satisfying the next condition for u' and ν.

CONDITION 8.10. (1) $[\Sigma',\vec{z}']$ is an element of the image of the source gluing map $\mathrm{Gluesoc}^{\mathbb{R}}_{\Xi^{\mathrm{ob}}}:\mathcal{V}_1^{\mathrm{ob},\mathbb{R}}\times\mathcal{V}_2^{\mathrm{ob},\mathbb{R}}\times[0,\nu)\to\mathcal{CM}_{g_1+g_2,\ell_1+\ell_2}$ associated with $\Xi^{\mathrm{ob}}=(\Xi_1^{\mathrm{ob}},\Xi_2^{\mathrm{ob}})$.
(2) For $z\in K_i^{\mathrm{ob}}$ we have

$$d(u_i^{\mathrm{ob}}(z),u'(\mathfrak{I}^{\sigma}_{\Xi^{\mathrm{ob}},i}(z)))\le\frac{\iota'_X}{2}.$$

Here ι'_X is the constant defined as in Condition 2.3 and $\mathfrak{I}^{\sigma}_{\Xi^{\mathrm{ob}},i}$ is the canonical embedding associated to Ξ^{ob}.

[2]Hereafter we abuse the notation and write an element of $\mathcal{M}_{g,k+1}$ as $(\Sigma,\vec{z})$ etc. in place of $[\Sigma,\vec{z}]$ etc. sometimes.

Now we define a map

$$I_{u',i} : \mathcal{E}_i^{\mathfrak{ob}} \to C^\infty(\Sigma'; (u')^*TX \otimes \Lambda^{0,1}(\Sigma')) \tag{8.19}$$

as follows. (See (3.14).) Let $z \in K_i^{\mathfrak{ob}}$. We have the parallel transport

$$(\mathrm{Pal}_{u_i^{\mathfrak{ob}}(z)}^{u'(z)})^J : T_{u_i^{\mathfrak{ob}}(z)}X \to T_{u'(z)}X, \tag{8.20}$$

defined in (2.5). On the other hand we have a projection

$$(\Lambda^{0,1}(\Sigma_i^{\mathfrak{ob}}))_z \to (\Lambda^{0,1}(\Sigma'))_z. \tag{8.21}$$

We remark that the map (8.21) is complex linear. The linear map $I_{u',i}$ is induced by the tensor product (over $\mathbb{C}$) of the two maps (8.20) and (8.21).

DEFINITION 8.11. Suppose $(\Sigma', \vec{z}')$ and u' satisfy Condition 8.10. We define the obstruction space $\mathcal{E}_i(u')$ at u' by

$$\mathcal{E}_i(u') := I_{u',i}(\mathcal{E}_i^{\mathfrak{ob}})$$

for the map $I_{u',i}$ in (8.19).

REMARK 8.12. We use a gluing datum centered at $(\Sigma_i^{\mathfrak{ob}}, \vec{z}_i^{\mathfrak{ob}})$ to define $I_{u',i}$ and $\mathcal{E}_i(u')$, instead of one centered at $(\Sigma_i, \vec{z}_i)$. This is an important point which enables us to define the coordinate change of Kuranishi structure. See Remark 8.26.

We now define the moduli spaces we study. Consider an obstruction bundle datum which is centered at $(((\Sigma_1^{\mathfrak{ob}}, \vec{z}_1^{\mathfrak{ob}}), u_1^{\mathfrak{ob}}), ((\Sigma_2^{\mathfrak{ob}}, \vec{z}_2^{\mathfrak{ob}}), u_2^{\mathfrak{ob}}))$, which we denote by $(\Xi^{\mathfrak{ob}}, (\mathcal{E}_1^{\mathfrak{ob}}, \mathcal{E}_2^{\mathfrak{ob}})$. We also consider $(((\Sigma_1, \vec{z}_1), u_1), ((\Sigma_2, \vec{z}_2), u_2))$ and a gluing datum Ξ centered at $((\Sigma_1, \vec{z}_1), (\Sigma_2, \vec{z}_2))$.

CONDITION 8.13. We assume that the pair $(((\Sigma_1, \vec{z}_1), u_1), ((\Sigma_2, \vec{z}_2), u_2))$ is close to the obstruction center $(((\Sigma_1^{\mathfrak{ob}}, \vec{z}_1^{\mathfrak{ob}}), u_1^{\mathfrak{ob}}), ((\Sigma_2^{\mathfrak{ob}}, \vec{z}_2^{\mathfrak{ob}}), u_2^{\mathfrak{ob}}))$ in the following sense. We also assume that the neighborhoods $\mathcal{V}_i^{\mathbb{R}}$ of $(\Sigma_i, \vec{z}_i)$ in $\mathcal{M}_{g_i,k_i+1}$ which is a part of data Ξ_i is small in the following sense.

(1) $(\Sigma_i, \vec{z}_i)$ is contained in the neighborhood $\mathcal{V}_i^{\mathfrak{ob},\mathbb{R}}$ of $(\Sigma_i^{\mathfrak{ob}}, \vec{z}_i^{\mathfrak{ob}})$ in $\mathcal{M}_{g_i,k_i+1}$ which is a part of $\Xi_i^{\mathfrak{ob}}$. Moreover $\mathcal{V}_i^{\mathbb{R}} \subset \mathcal{V}_i^{\mathfrak{ob},\mathbb{R}}$

(2) Let $(\Sigma_i, \vec{z}_i) \cong (\Sigma_i^{\mathfrak{ob}}(\sigma_i^0), \vec{z}_i^{\mathfrak{ob}}(\sigma_i^0))$ with $\sigma_i^0 \in \mathcal{V}_i^{\mathfrak{ob}.\mathbb{R}}$. The gluing datum $\Xi_i^{\mathfrak{ob}}$ determines a diffeomorphism

$$\mathfrak{I}_i^0 : \Sigma_i^{\mathfrak{ob}} \to \Sigma_i^{\mathfrak{ob}}(\sigma_i^0) \cong \Sigma_i.$$

(Here the first diffeomorphism is induced by $\Xi_i^{\mathfrak{ob}}$ and the second diffeomorphism is the identification $[\Sigma_i, \vec{z}_i] = [\Sigma_i^{\mathfrak{ob}}(\sigma_i^0),\ \vec{z}_i^{\mathfrak{ob}}(\sigma_i^0)]$.) We require

$$\sup\{d(u_i(\mathfrak{I}_i^0(z)), u_i^{\mathfrak{ob}}(z)) \mid z \in \Sigma_i^{\mathfrak{ob}}, i = 1, 2\} \le \frac{\iota'_X}{4}. \tag{8.22}$$

(3) We also require

$$\mathfrak{I}_i^0(K_i^{\mathfrak{ob}}) \subset \mathrm{Int}\, K_i. \tag{8.23}$$

We recall that $(\Sigma_\infty, \vec{z}_\infty)$ is a union of $(\Sigma_1, \vec{z}_1)$ and $(\Sigma_2, \vec{z}_2)$, which are glued to each other at their 0-th marked points, which may also carry their marked points other than the 0-th ones.

DEFINITION 8.14. We assume that $(((\Sigma_1^{\mathfrak{ob}},\vec z_1^{\mathfrak{ob}}),u_1^{\mathfrak{ob}}),((\Sigma_2^{\mathfrak{ob}},\vec z_2^{\mathfrak{ob}}),u_2^{\mathfrak{ob}}))$, $(\Xi^{\mathfrak{ob}},(\mathcal{E}_1^{\mathfrak{ob}},\mathcal{E}_2^{\mathfrak{ob}}))$, $(((\Sigma_1,\vec z_1),u_1),((\Sigma_2,\vec z_2),u_2))$ and Ξ satisfy Condition 8.13 and $(\epsilon,\nu)\in\mathbb{R}_+^2$.

We define the moduli space $\mathcal{M}_+^{\mathcal{E}_1\oplus\mathcal{E}_2}((\Sigma_\infty,\vec z);u_1,u_2)_{\epsilon,\nu}$ as the set of all the isomorphism classes of $((\Sigma',\vec z'),u')$ such that the following three conditions are satisfied.

(1) $(\Sigma',\vec z') = \mathrm{Gluesoc}_\Xi^{\mathbb{R}}(\sigma_1,\sigma_2,T)\in \mathrm{Im}(\mathrm{Gluesoc}_\Xi^{\mathbb{R}})$. (σ_1,σ_2) is in the ϵ neighborhood of σ_0 in $\mathcal{V}_1^{\mathbb{R}}\times\mathcal{V}_2^{\mathbb{R}}$. ($\mathcal{V}_1^{\mathbb{R}}$ and $\mathcal{V}_2^{\mathbb{R}}$ are parts of Ξ.) Here σ_0 corresponds to $(\Sigma_1,\vec z_1)$ and $(\Sigma_2,\vec z_2)$ and $T>1/\nu$.

(2) Condition 3.7 is satisfied. Namely:
   (a) We assume $u_i|_{K_i}$ is ϵ close to $u'\circ\mathfrak{I}_{\Xi,i}^{\sigma}|_{K_i}$ in C^1 sense. Here $\mathfrak{I}_{\Xi,i}^{\sigma}$ is defined in Definition-Lemma 8.7.
   (b) $\mathrm{Diam}\{u'(z)\mid z\in\Sigma'\setminus(K_1\cup K_2)\}<\epsilon$.

(3)

$$\overline\partial u' \equiv 0 \mod \mathcal{E}_1(u')\oplus\mathcal{E}_2(u'). \tag{8.24}$$

Note (1), (2) and (8.19), (8.22) imply that $\mathcal{E}_1(u')$, $\mathcal{E}_2(u')$ are defined if ϵ is sufficiently small.

Note that if $(\Sigma_i^{\mathfrak{ob}},\partial\Sigma_i^{\mathfrak{ob}})\cong(\Sigma_i,\partial\Sigma_i)$ as bordered Riemann surfaces and $\Xi_{\mathfrak{ob}}=\Xi$, the moduli space $\mathcal{M}^{\mathcal{E}_1\oplus\mathcal{E}_2}((\Sigma_T,\vec z);(u_1,u_2))_\epsilon$ we defined in Definition 3.8 is a subset of the moduli space $\mathcal{M}_+^{\mathcal{E}_1\oplus\mathcal{E}_2}((\Sigma_\infty,\vec z_\infty);u_1,u_2)_{\epsilon,\nu}$ we defined here, for $T>1/\nu$.

We also define

$$I_{u_i'}:\mathcal{E}_i^{\mathfrak{ob}}\to\Gamma(\Sigma_i;(u_i')^*TX\otimes\Lambda^{0,1}(\Sigma_i)) \tag{8.25}$$

as in (3.20). Namely it is induced from the tensor product over $\mathbb{C}$ of the projection $(\Lambda^{0,1}(\Sigma_i^{\mathfrak{ob}}))_z\to(\Lambda^{0,1}(\Sigma_i(\sigma_i)))_z$, and the complex linear part of the parallel transport

$$\left(\mathrm{Pal}_{u_i^{\mathfrak{ob}}(z)}^{u_i'(z)}\right)^J : T_{u^{\mathfrak{ob}}(z)}X\to T_{u_i'(z)}X.$$

We put

$$\mathcal{E}_i(u_i') = I_{u_i'}(\mathcal{E}_i^{\mathfrak{ob}}). \tag{8.26}$$

So we can define an equation

$$\overline\partial u_i'\equiv 0,\qquad \mod \mathcal{E}_i(u_i'). \tag{8.27}$$

DEFINITION 8.15. $\mathcal{M}_+^{\mathcal{E}_i}((\Sigma_i,\vec z_i);u_i)_\epsilon$ is the set of isomorphism classes of $((\Sigma_i',\vec z_i'),u_i')$ with the following properties.

(1) $(\Sigma_i',\vec z_i')$ represents an element σ_i of $\mathcal{V}_i^{\mathbb{R}}$ and is in an ϵ neighborhood of $[\Sigma_i,\vec z_i]$.
(2) u_i' is ϵ close to u_i in C^1 topology.
(3) u_i' satisfies equation (8.27).

We assumed Assumption 3.12 in Definition 8.9 (4), where $\mathrm{ev}_{i,\infty}$ in (3.24), (3.25) are ev_0 here. Then we can take ϵ small so that $\mathcal{M}_+^{\mathcal{E}_i}((\Sigma_i,\vec z_i);u_i)_\epsilon$ is a smooth manifold. Moreover the fiber product

$$\mathcal{M}_+^{\mathcal{E}_1}((\Sigma_1,\vec z_1);u_1)_{\epsilon\ \mathrm{ev}_0}\times_{\mathrm{ev}_0}\mathcal{M}_+^{\mathcal{E}_2}((\Sigma_2,\vec z_2);u_2)_\epsilon \tag{8.28}$$

is transversal. (Here the fiber product is taken with respect to the evaluation maps $\mathrm{ev}_0 = \mathrm{ev}_{i,\infty}$ $(i = 1, 2)$ given in Chapter 3.)

Now Theorems 3.13 and 6.4 are generalized as follows.

THEOREM 8.16. *For any* $\epsilon(11), \nu(1) > 0$ *there exist* $\epsilon(12) > 0$, $T_{7,m,\epsilon(11),\nu(1)} > 0$ *and a map*

$$\mathrm{Glue}_+ : \mathcal{M}_+^{\mathcal{E}_1}((\Sigma_1, \vec{z}_1); u_1)_{\epsilon(12)} \ {}_{\mathrm{ev}_0}\times_{\mathrm{ev}_0} \mathcal{M}_+^{\mathcal{E}_2}((\Sigma_2, \vec{z}_2); u_2)_{\epsilon(12)} \times (T_{7,m,\epsilon(11),\nu(1)}, \infty] \\ \to \mathcal{M}_+^{\mathcal{E}_1 \oplus \mathcal{E}_2}((\Sigma_\infty, \vec{z}); u_1, u_2)_{\epsilon(11),\nu(1)}$$

with the following properties.

(1) *The map* Glue_+ *is a homeomorphism onto its image. The image contains* $\mathcal{M}_+^{\mathcal{E}_1 \oplus \mathcal{E}_2}((\Sigma_\infty, \vec{z}); u_1, u_2)_{\epsilon_{\epsilon(11),\nu(1),(8.29)}, \nu_{\epsilon(11),\nu(1),(8.29)}}$, *where* $\epsilon_{\epsilon(11),\nu(1),(8.29)}$, *and* $\nu_{\epsilon(11),\nu(1),(8.29)}$ *are positive numbers depending on* $\epsilon(11), \nu(1)$.

(2) *The next diagram commutes*

$$(8.29) \qquad \begin{array}{ccc} \begin{array}{c}\mathcal{M}_+^{\mathcal{E}_1}((\Sigma_1,\vec{z}_1);u_1)_{\epsilon(12)} \ {}_{\mathrm{ev}_0}\times_{\mathrm{ev}_0} \\ \mathcal{M}_+^{\mathcal{E}_2}((\Sigma_2,\vec{z}_2);u_2)_{\epsilon(12)} \times (T_{7,m,\epsilon(11),\nu(1)},\infty]\end{array} & \longrightarrow & \mathcal{V}_1^{\mathbb{R}} \times \mathcal{V}_2^{\mathbb{R}} \times (T_{7,m,\epsilon(11),\nu(1)}, \infty] \\ \Big\downarrow {\scriptstyle \mathrm{Glue}_+} & & \Big\downarrow {\scriptstyle \mathrm{Gluesoc}_{\Xi}^{\mathbb{R}}} \\ \mathcal{M}_+^{\mathcal{E}_1 \oplus \mathcal{E}_2}((\Sigma_\infty, \vec{z}_\infty); u_1, u_2)_{\epsilon(11),\nu(1)} & \longrightarrow & \mathcal{CM}_{g_1+g_2,\ell_1+\ell_2} \end{array}$$

where the horizontal arrows are defined by forgetting the map part.

(3) *The map* Glue_+ *defines a fiberwise diffeomorphism of* C^{m-2} *class onto its image. Here fiber means the fiber of the horizontal arrows of* (8.29).

This is the family version of Theorem 3.13 incorporating the variation of complex structures on the source curve into the gluing. A generalization of Theorem 6.4 is the following Theorem 8.17. We recall the definition

$$K_1^S = K_1 \cup ([0, S]_{\tau'} \times [0, 1]), \quad K_2^S = K_2 \cup ([-S, 0]_{\tau''} \times [0, 1])$$

from (6.4). We define

$$\mathrm{Glures}_{i,+,S} : \mathcal{M}_+^{\mathcal{E}_1}((\Sigma_1, \vec{z}_1); u_1)_{\epsilon(12)} \ {}_{\mathrm{ev}_0}\times_{\mathrm{ev}_0} \mathcal{M}_+^{\mathcal{E}_2}((\Sigma_2, \vec{z}_2); u_2)_{\epsilon(12)} \\ \times (T_{7,m,\epsilon(11),\nu(1)}, \infty] \to \mathrm{Map}_{L^2_{m+1}}((K_i^S, K_i^S \cap \partial\Sigma_i), (X, L))$$

as the composition of Glue_+ with the restriction map

$$\mathcal{M}_+^{\mathcal{E}_1 \oplus \mathcal{E}_2}((\Sigma_\infty, \vec{z}); u_1, u_2)_{\epsilon(11),\nu(1)} \to \mathrm{Map}_{L^2_{m+1}}((K_i^S, K_i^S \cap \partial\Sigma_i), (X, L)).$$

Here we use the map $\mathfrak{I}^{\sigma}_{\Xi,i}$ in Lemma-Definition 8.7 to make the identification

$$K_i^S \cong K_i^S(\sigma_i) \subset \Sigma_1(\sigma_1) \#_T \Sigma_2(\sigma_2).$$

Note the diffeomorphism $K_i^S \cong K_i^S(\sigma_i)$ is induced by the gluing datum Ξ centered at $((\Sigma_1, \vec{z}_1), (\Sigma_2, \vec{z}_2))$.

The following exponential estimate is the crux entering in the proof of smoothness of the coordinate change of the Kuranishi chart.

THEOREM 8.17. *There exists* $\delta_2 > 0$ *with the following property. For each* m *and* S, *there exist constants* $T_{m,S,(8.30)} > \frac{S}{10}$ *and* $C_{m,S,(8.30)} > 0$ *such that*

$$(8.30) \qquad \left\| \nabla^n_{\sigma,\rho} \frac{d^\ell}{dT^\ell} \mathrm{Glures}_{i,+,S} \right\|_{L^2_{m+1-\ell}} < C_{m,S,(8.30)} e^{-\delta_2 T}$$

for all $m \geq n \geq 0$, $m \geq \ell > 0$. *Here* $\nabla_{\sigma,\rho}$ *is a differentiation with respect to* $((\sigma_1, \rho_1), (\sigma_2, \rho_2))$ *in the fiber product* (8.28). [3]

PROOF OF THEOREMS 8.16 AND 8.17. We denote by $u_i^{\rho_i} : \Sigma_i(\sigma_i) \to X$ the map corresponding to $(\sigma_i, \rho_i) \in \mathcal{M}_+^{\mathcal{E}_i}((\Sigma_i, \vec{z}_i); u_i)_{\epsilon(12)}$.

For each

$$((\sigma_1, \rho_1), (\sigma_2, \rho_2)) \in \mathcal{M}_+^{\mathcal{E}_1}((\Sigma_1, \vec{z}_1); u_1)_{\epsilon(12)} \;{}_{\mathrm{ev}_0}\times_{\mathrm{ev}_0} \mathcal{M}_+^{\mathcal{E}_2}((\Sigma_2, \vec{z}_2); u_2)_{\epsilon(12)}$$

and T we run the alternating method developed in Chapters 5 and 6.

Namely we start with $((\Sigma_1(\sigma_1), \vec{z}_1(\sigma_1)), u_1^{\rho_1}), (\Sigma_2(\sigma_2), \vec{z}_1(\sigma_2)), u_2^{\rho_2})$ in place of $((\Sigma_1, \vec{z}_1), u_1), ((\Sigma_2, \vec{z}_2), u_2)$ and use the coordinates at 0-th marked points which we determined as a part of the gluing data Ξ centered at $((\Sigma_1, \vec{z}_1), u_1), ((\Sigma_2, \vec{z}_2), u_2))$.

Except the point which we will explain below, the proof goes in the same way as before and we obtain a map

$$V_1(\sigma_1, \epsilon) \times_L V_2(\sigma_2, \epsilon) \times (T_0, \infty] \to \mathcal{M}_+^{\mathcal{E}_1 \oplus \mathcal{E}_2}((\Sigma_\infty, \vec{z}); u_1, u_2)_{\epsilon,\nu}.$$

Here $V_i(\sigma_i, \epsilon)$ is the fiber of the map $\mathcal{M}_+^{\mathcal{E}_i}((\Sigma_i, \vec{z}_i); u_i)_\epsilon \to \mathcal{V}_i^{\mathbb{R}}$ at $\sigma_i \in \mathcal{V}_i^{\mathbb{R}}$ and $T_0 > 1/\nu$. The union of these maps over σ_1, σ_2 will be the map Glue_+.

The point we need to clarify in adapting the proof of Chapters 5 and 6 to our situation is the following: The way we define $\mathcal{E}_i(u')$ in this chapter is slightly different from that of Chapters 5 and 6. Namely, in this chapter, we start with $\mathcal{E}_i^{\mathfrak{ob}}$ defined on $(\Sigma_i^{\mathfrak{ob}}, u_i^{\mathfrak{ob}})$ with $\Sigma_i^{\mathfrak{ob}} \neq \Sigma_i$, while in Chapters 5 and 6 $\mathcal{E}_i^{\mathfrak{ob}}$ was defined on $(\Sigma_i, u_i^{\mathfrak{ob}})$, i.e., $\Sigma_i^{\mathfrak{ob}} = \Sigma_i$.

We use Proposition 8.19 below to obtain the required estimate for the current $\mathcal{E}_i(u')$. Then the arguments in Chapters 5 and 6 also go through for the proofs of Theorems 8.16 and 8.17.

In fact Theorem 8.16 (1),(3) are proved in the same way as in Chapters 5, 6 and 7. The estimate of σ derivative is the same as that of ρ derivative. So the proof of Theorem 8.17 is the same as that in Chapter 6. Theorem 8.16 (2) follows from the fact that the alternating method we use does not change the complex structure of the source.

To formulate Proposition 8.19, we need some preparation.

Let $(\Sigma', \vec{z}') = \mathrm{Gluesoc}_{\Xi}^{\mathbb{R}}((\sigma_1, \sigma_2), T)$ and $u' : (\Sigma', \partial\Sigma') \to (X, L)$ a smooth map satisfying the conditions

$$\begin{aligned} d(u'(z), u_i(z)) &\leq \epsilon, \qquad \text{for } z \in K_i \\ \mathrm{Diam}(u'(\Sigma' \setminus (K_1 \cup K_2))) &\leq \epsilon. \end{aligned} \tag{8.31}$$

We take a basis $\mathbf{e}_{i,a}$ $(a = 1, \dots, \dim \mathcal{E}_i^{\mathfrak{ob}})$ of $\mathcal{E}_i^{\mathfrak{ob}} \subset \Gamma(K_i^{\mathfrak{ob}}; (u_i^{\mathfrak{ob}})^* TX \otimes \Lambda^{0,1}(\Sigma_i^{\mathfrak{ob}}))$ and put

$$\mathbf{e}_{i,a}^0(u') = (I_{u_i^{\rho_i}}^{u_i} \circ I_{u'}^{u_i^{\rho_i}} \circ I_{u',i})(\mathbf{e}_{i,a}) \in \Gamma(K_i; (u_i)^* TX \otimes \Lambda^{0,1}(\Sigma_i)). \tag{8.32}$$

Here the map $I_{u',i}$ is (8.19) and the maps

$$\begin{aligned} I_{u'}^{u_i^{\rho_i}} &: \Gamma(\mathfrak{I}_{\Xi^{\mathfrak{ob}},i}^{\sigma'}(K_i^{\mathfrak{ob}}); (u')^* TX \otimes \Lambda^{0,1}(\Sigma')) \to \Gamma(\Sigma_i(\sigma_i); (u_i^{\rho_i})^* TX \otimes \Lambda^{0,1}(\Sigma_i(\sigma_i))) \\ I_{u_i^{\rho_i}}^{u_i} &: \Gamma(\Sigma_i(\sigma_i); (u_i^{\rho_i})^* TX \otimes \Lambda^{0,1}(\Sigma_i(\sigma_i))) \to \Gamma(K_i; (u_i)^* TX \otimes \Lambda^{0,1}(\Sigma_i)) \end{aligned}$$

---

[3] σ_1 and σ_2 parameterize the source curve and ρ_1 and ρ_2 parameterize the map.

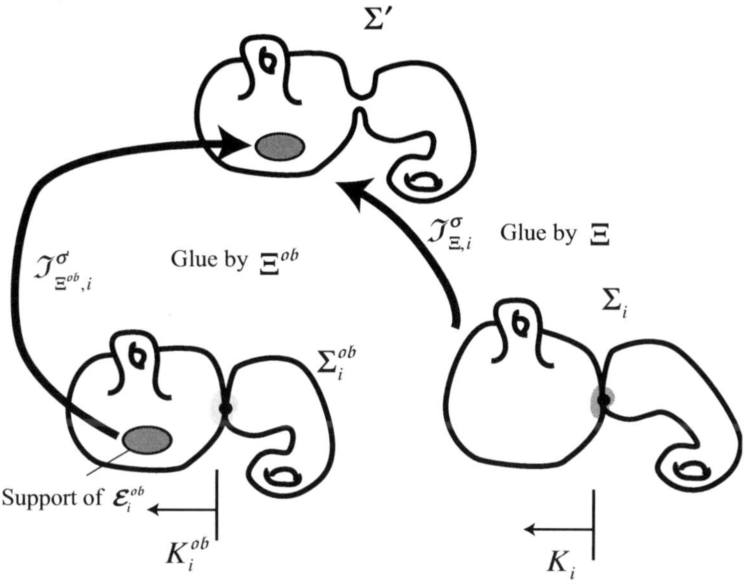

FIGURE 8.4. Two gluing data give different embeddings.

are defined as follows. Note since $(\Sigma', \vec{z}') = \mathrm{Gluesoc}^{\mathbb{R}}_{\Xi^{\mathfrak{ob}}}((\sigma_1', \sigma_2'), T')$ we have a smooth embedding

$$\mathfrak{I}^{\sigma'}_{\Xi^{\mathfrak{ob}},i} : K_i^{\mathfrak{ob}} \to \Sigma' \tag{8.33}$$

by Definition-Lemma 8.7. Its image appears in the domain of $I^{u_i^{\rho_i}}_{u'}$.

We use an embedding

$$\mathfrak{I}^{\sigma}_{\Xi,i} : K_i \to \Sigma' \tag{8.34}$$

which is also obtained by Definition-Lemma 8.7 in the definition of $I^{u_i^{\rho_i}}_{u'}$.

REMARK 8.18. We remark that in (8.33) we use the gluing data $\Xi^{\mathfrak{ob}}$ centered at $((\Sigma_1^{\mathfrak{ob}}, \vec{z}_1^{\mathfrak{ob}}), (\Sigma_2^{\mathfrak{ob}}, \vec{z}_2^{\mathfrak{ob}}))$. In (8.34) we use the gluing data $\Xi = (\Xi_1, \Xi_2)$ centered at $((\Sigma_1, \vec{z}_1), (\Sigma_2, \vec{z}_2))$. See Figure 8.4.

We assume the next relation:

$$\mathfrak{I}^{\sigma'}_{\Xi^{\mathfrak{ob}},i}(K_i^{\mathfrak{ob}}) \subset \mathfrak{I}^{\sigma}_{\Xi,i}(K_i) \tag{8.35}$$

which is needed so that the above composition is defined. By Condition 8.13 we may choose $\mathcal{V}_i^{\mathbb{R}}$, a neighborhood of $[\Sigma_i, \vec{z}_i]$ in the moduli space of marked bordered curves, so small that if $(\sigma_1, \sigma_2) \in \mathcal{V}_1^{\mathbb{R}} \times \mathcal{V}_2^{\mathbb{R}}$ then (8.35) is satisfied.

The smooth embedding $\mathfrak{I}^{\sigma}_{\Xi,i}$ induces a complex linear map

$$\Lambda^{0,1}_z(\Sigma') \to \Lambda^1_{\mathfrak{I}^{\sigma}_{\Xi,i}(z)}(\Sigma_i(\sigma_i)) \to \Lambda^{0,1}_{\mathfrak{I}^{\sigma}_{\Xi,i}(z)}(\Sigma_i(\sigma_i)). \tag{8.36}$$

Here the second map is the projection and the first map is the complex linear part of the map induced by $\mathfrak{I}^{\sigma}_{\Xi,i}$. Namely (8.36) is given by

$$\frac{1}{2}\left(((\mathfrak{I}^{\sigma}_{\Xi,i})^{-1})^* - j_{\Sigma_i(\sigma_i)} \circ ((\mathfrak{I}^{\sigma}_{\Xi,i})^{-1})^* \circ j_{\Sigma^{\mathfrak{ob}}}\right).$$

We also consider the map:

$$\left(\left(\mathrm{Pal}_{u_i^{\rho_i}(z)}^{u'(z)}\right)^J\right)^{-1} : T_{u'(z)}X \to T_{u_i^{\rho_i}(z)}X. \tag{8.37}$$

Then $I_{u'}^{u^{\rho_i}}$ is obtained from the tensor product of (8.36) and (8.37).

Similarly $I_{u_i^{\rho_i}}^{u_i}$ is induced by the tensor product of the projection

$$\Lambda_z^{0,1}(\Sigma_i(\sigma_i)) \to \Lambda_z^{0,1}(\Sigma_i) \tag{8.38}$$

and a complex linear part of the parallel transformation

$$\left(\left(\mathrm{Pal}_{u_i(z)}^{u_i^{\rho_i}(z)}\right)^J\right)^{-1} : T_{u_i^{\rho_i}(z)}X \to T_{u_i(z)}X. \tag{8.39}$$

This finishes the explanation of the maps (8.32).

Let

$$(\sigma,\rho) = ((\sigma_1,\rho_1),(\sigma_2,\rho_2)) \in \mathcal{M}_+^{\mathcal{E}_1}((\Sigma_1,\vec{z}_1);u_1)_\epsilon \times_L \mathcal{M}_+^{\mathcal{E}_2}((\Sigma_2,\vec{z}_2);u_2)_\epsilon$$

and $(\Sigma_T^\sigma, \vec{z}_T^\sigma) = \mathrm{Gluesoc}^{\mathbb{R}}(\sigma_1,\sigma_2,T)$.

We denote by

$$\hat{u}_{i,T,(\kappa)}^{\sigma,\rho} : \Sigma_i^\sigma \cong \Sigma_i \to X$$

the map obtained at the κ-th inductive step of our inductive construction starting from $((\Sigma_1(\sigma_1),\vec{z}_1(\sigma_1)),u_1^{\rho_1})$, $((\Sigma_2(\sigma_2),\vec{z}_2(\sigma_2)),u_2^{\rho_2})$. The diffeomorphism $\Sigma_i^\sigma \cong \Sigma_i$ is a part of datum Ξ_i.

In other words $\hat{u}_{i,T,(\kappa)}^{\sigma,\rho}$ corresponds to the map $\hat{u}_{i,T,(\kappa)}^{\rho}$ in (5.74). We put

$$\mathbf{e}'_{i,a;(\kappa)}(\sigma,\rho,T) = \mathbf{e}_{i,a}^0(u_{T,(\kappa)}^{\sigma,\rho}) \in \Gamma(K_i; u_i^*TX \otimes \Lambda^{0,1}(\Sigma_i)). \tag{8.40}$$

Here $\mathbf{e}_{i,a}^0(u_{T,(\kappa)}^{\sigma,\rho})$ is the map defined as in (8.32), and the map $u_{T,(\kappa)}^{\sigma,\rho}$ is the ρ-parametrized version of the map $u_{T,(\kappa)}^{\sigma}$ in Definition 5.33. We remark that the set $\{\mathbf{e}'_{i,a;(\kappa)}(\sigma,\rho,T)\}$ forms a basis of the subspace

$$\left(I_{u_i^{\rho_i}}^{u_i} \circ I_{\hat{u}_{i,T,(\kappa)}^{\sigma,\rho}}^{u_i^{\rho_i}}\right)(\mathcal{E}_i(\hat{u}_{i,T,(\kappa)}^{\sigma,\rho})).$$

Moreover there exists a canonical isomorphism

$$\mathcal{E}_i(\hat{u}_{i,T,(\kappa)}^{\sigma,\rho}) \cong \mathcal{E}_i(u_{T,(\kappa)}^{\sigma,\rho}). \tag{8.41}$$

(8.41) is a consequence of the equality $\hat{u}_{i,T,(\kappa)}^{\sigma,\rho} = u_{T,(\kappa)}^{\sigma,\rho}$ on $K_i \subset \Sigma_i(\sigma_i)$ which we regard as a subset $K_i \subset \Sigma_T^\sigma$ by the canonical embedding $\mathfrak{I}_{\Xi,i}^\sigma$. This equality follows from the definition (5.74).

PROPOSITION 8.19. *There exists $\delta_3 > 0$ such that we can estimate $\mathbf{e}'_{i,a;(\kappa)}(\sigma,\rho,T)$ and its T, ρ, σ derivatives as*

$$\left\| \nabla_{\sigma,\rho}^n \frac{\partial^\ell}{\partial T^\ell} \mathbf{e}'_{i,a;(\kappa)}(\sigma,\rho,T) \right\|_{L^2_{m+1-\ell}} \le C_{m,(8.42)} e^{-\delta_3 T} \tag{8.42}$$

for $m-2 \ge n \ge 0$, $m-2 \ge \ell > 0$, if $T > T_{m,(8.42)}$.

This proposition is the family version of Lemma 6.21.[4] The proof will be given in Section 8.4.

[4] Actually Lemma 6.21 gives an estimate of the orthonormal frame obtained from $\mathbf{e}'_{i,a;(\kappa)}(\rho,T)$ by the Gram-Schmidt process. However the argument to study the Gram-Schmidt process in our situation is the same as the one in the proof of Lemma 6.21.

We can use Proposition 8.19 to control the obstruction bundle $\mathcal{E}_i(u')$ together with its derivatives.

More explicitly: we can use the proposition to show (5.46): we use it to show a version of Lemma G.2, where σ derivative as well as ρ and T derivatives is included and the definition of $\mathcal{E}_i(u')$ is slightly modified as we mentioned at the beginning the proof of Theorem 8.17.

We can use Proposition 8.19 also in all the other places where we need to control $\mathcal{E}_i(u')$. Note we take $\delta > 0$ in Definition 3.4 so that not only $\delta < \delta_1/10$ but also $\delta < \delta_3/10$ holds. The constant δ_2 in Theorem 8.17 is determined by δ_1 and δ_3.

The proofs of Theorems 8.16 and 8.17 are now complete except the proof of Proposition 8.19. □

8.3. Smoothness of coordinate change

We now use Theorems 8.16 and 8.17 to prove smoothness of the coordinate change. We begin with explaining the situation where we study the coordinate change.

SITUATION 8.20. Let $(\Sigma_i^{\mathfrak{ob}}, \vec{z}_i^{\mathfrak{ob}}) \in \mathcal{M}_{g_i,k_i+1}$ for $i = 1,2$. We take $\Xi^{\mathfrak{ob}} = (\Xi_1^{\mathfrak{ob}}, \Xi_2^{\mathfrak{ob}})$ a gluing datum centered at $((\Sigma_1^{\mathfrak{ob}}, \vec{z}_1^{\mathfrak{ob}}), (\Sigma_2^{\mathfrak{ob}}, \vec{z}_2^{\mathfrak{ob}}))$. We consider $u_i^{\mathfrak{ob}} : (\Sigma_i^{\mathfrak{ob}}, \partial\Sigma_i^{\mathfrak{ob}}) \to (X, L)$, a pseudoholomorphic map of homology class β_i. For $j = 1, 2$, let $(\Xi^{\mathfrak{ob}}, (\mathcal{E}_1^{(j)}, \mathcal{E}_2^{(j)}))$ be an obstruction bundle datum which is centered at $(((\Sigma_1^{\mathfrak{ob}}, \vec{z}_1^{\mathfrak{ob}}), u_1^{\mathfrak{ob}}), ((\Sigma_2^{\mathfrak{ob}}, \vec{z}_2^{\mathfrak{ob}}), u_2^{\mathfrak{ob}}))$. Note we use the same gluing datum $\Xi^{\mathfrak{ob}}$ for $j = 1, 2$.

We assume that the next inclusion holds for $i = 1, 2$.

$$\mathcal{E}_i^{(1)} \subseteq \mathcal{E}_i^{(2)}. \tag{8.43}$$

In sum we take two obstruction bundle data with the same gluing dataum that satisfies (8.43).

SITUATION 8.21. We next consider $((\Sigma_i^{(j)}, \vec{z}_i^{(j)}), u_i^{(j)})$ for $i, j = 1, 2$. Here:

(1) $(\Sigma_i^{(j)}, \vec{z}_i^{(j)}) \in \mathcal{M}_{g_i,k_i+1}$.
(2) For each $i = 1, 2$, $j = 1, 2$, $u_i^{(j)} : (\Sigma_i^{(j)}, \partial\Sigma_i^{(j)}) \to (X, L)$ is a pseudoholomorphic map of homology class β_i.

We call such a pair $((\Sigma_1^{(j)}, \vec{z}_1^{(j)}), u_1^{(j)})), ((\Sigma_2^{(j)}, \vec{z}_2^{(j)}), u_2^{(j)}))$ for $j = 1$ or 2, a *chart center*.

Let $\Xi^{(j)} = (\Xi_1^{(j)}, \Xi_2^{(j)})$ be a gluing datum centered at $(\Sigma_i^{(j)}, \vec{z}_i^{(j)})$. We assume

$$\mathcal{V}_i^{(1),\mathbb{R}} \subset \mathcal{V}_i^{(2),\mathbb{R}} \tag{8.44}$$

where $\mathcal{V}_i^{(j),\mathbb{R}}$ is a neighborhood of $(\Sigma_i^{(j)}, \vec{z}_i^{(j)})$ in $\mathcal{M}_{g_i,k_i+1}$ which is a part of $\Xi_i^{(j)}$. Note (8.44) implies $(\Sigma_i^{(1)}, \vec{z}_i^{(1)}) \in \mathcal{V}_i^{(2),\mathbb{R}}$.

We also assume Conditions 8.22 and 8.23 below.

When $u_i : (\Sigma_i, \partial\Sigma_i) \to (X, L)$ is given and satisfies $\mathrm{ev}_{1,\infty}(u_1) = \mathrm{ev}_{1,\infty}(u_2)$, we denote by $u_\infty : (\Sigma_\infty, \partial\Sigma_\infty) \to (X, L)$ the map that restricts to u_i on Σ_i.

We take and fix $\epsilon_6, \nu_1 > 0$. We put $\epsilon(11) = \epsilon_6$ and $\nu(1) = \nu_1$ in Theorem 8.16. (Here we will apply Theorem 8.16 to $\mathrm{Glue}_+^{(2)}$ in (8.49).) We put

$$\epsilon_7 = \epsilon_{\epsilon(11),\nu(1),(8.29)}, \qquad \nu_2 = \nu_{\epsilon(11),\nu(1),(8.29)}, \tag{8.45}$$

where the right hand sides are obtained in Theorem 8.16.

CONDITION 8.22. We identify $(\Sigma_1^{(j)}, \vec{z}_1^{(j)})$ with $(\Sigma_2^{(j)}, \vec{z}_2^{(j)})$ at their 0-th marked points to obtain $(\Sigma_\infty^{(j)}, \vec{z}^{(j)})$. Together with $u_1^{(j)}$ and $u_2^{(j)}$ it gives an element of $\mathcal{M}_{g_1+g_2,k_1+k_2}(X, L; \beta)$ which we denote by $[(\Sigma_\infty^{(j)}, \vec{z}^{(j)}), u_\infty^{(j)}]$. We require

$$((\Sigma_\infty^{(1)}, \vec{z}^{(1)}), u_\infty^{(1)}) \in \mathcal{M}_+^{\mathcal{E}_1^{(2)} \oplus \mathcal{E}_2^{(2)}}((\Sigma_\infty^{(2)}, \vec{z}^{(2)}); u_1^{(2)}, u_2^{(2)})_{\epsilon_7/2, \nu_2/2}. \tag{8.46}$$

Roughly speaking Condition 8.22 means that $u_i^{(1)}$ is close to $u_i^{(2)}$.

CONDITION 8.23. Let $\sigma = (\sigma_1, \sigma_2) \in \mathcal{V}_1^{(1),\mathbb{R}} \times \mathcal{V}_2^{(1),\mathbb{R}}$. By (8.44) $\sigma \in \mathcal{V}_1^{(2),\mathbb{R}} \times \mathcal{V}_2^{(2),\mathbb{R}}$. Using the gluing data $\Xi^{(1)}$, $\Xi^{(2)}$ taken in Situation 8.21, we obtain a σ_i dependent family of embeddings

$$\psi_{\sigma_i} : K_i^{(2)} \subset \Sigma_i^{(2)} \cong \Sigma_i^{(2)}(\sigma_i) \cong \Sigma_i^{(1)}(\sigma_i), \tag{8.47}$$

for $i = 1, 2$, where the first inclusion is given by definition, the first $\cong$ is the σ_i dependent family of diffeomorphisms which is a part of the datum $\Xi^{(2)}$ and the second $\cong$ is the unique biholomorphic map.

We require

$$\psi_{\sigma_i}(K_i^{(2)}) \subseteq \operatorname{Int} K_i^{(1)} \tag{8.48}$$

for $\sigma_i \in \mathcal{V}_i^{(2),\mathbb{R}}$.

Note that for given $\Xi^{(1)}$, $\Xi^{(2)}$ we can always modify the analytic family $\varphi_{i,(2)}^{\mathbb{R}}$ of coordinates, which is a part of $\Xi^{(2)}$ so that Condition 8.23 is satisfied. In fact we may replace $\varphi_{i,(2)}^{\mathbb{R}}$ by its conformal change $z \mapsto \varphi_{i,(2)}^{\mathbb{R}}(\epsilon z)$ for sufficiently small ϵ. In other words we may assume Condition 8.23 without loss of generality.

Theorem 8.25, the main result of this chapter, concerns the first vertical arrow of the next Diagram (8.49).

LEMMA 8.24. *There exist constants ϵ_8, ϵ_9, $\epsilon_{10,m}$, $\nu_{3,m}$, $\nu_{4,m}$, and $T_{m,(8.49)}$ such that we have the following commutative diagram for any positive numbers $\epsilon \le \epsilon_{10,m}$ and $T(0) \ge T_{m,(8.49)}$.*

(8.49)

$$\begin{array}{ccc}
\begin{array}{c}\mathcal{M}_+^{\mathcal{E}_1^{(1)}}((\Sigma_1^{(1)},\vec{z}_1^{(1)});u_1^{(1)})_{\epsilon}\ {}_{\mathrm{ev}_0}\!\times_{\mathrm{ev}_0} \\ \mathcal{M}_+^{\mathcal{E}_2^{(1)}}((\Sigma_2^{(1)},\vec{z}_2^{(1)});u_2^{(1)})_{\epsilon}\times(T(0),\infty]\end{array} & \xrightarrow{\mathrm{Glue}_+^{(1)}} & \mathcal{M}_+^{\mathcal{E}_1^{(1)}\oplus\mathcal{E}_2^{(1)}}((\Sigma_\infty^{(1)},\vec{z}^{(1)});u_1^{(1)},u_2^{(1)})_{\epsilon_9,\nu_{3,m}} \\
\Big\downarrow \mathfrak{F} & & \Big\downarrow \\
\begin{array}{c}\mathcal{M}_+^{\mathcal{E}_1^{(2)}}((\Sigma_1^{(2)},\vec{z}_1^{(2)});u_1^{(2)})_{\epsilon_8}\ {}_{\mathrm{ev}_0}\!\times_{\mathrm{ev}_0} \\ \mathcal{M}_+^{\mathcal{E}_2^{(2)}}((\Sigma_2^{(2)},\vec{z}_2^{(2)});u_2^{(2)};\beta_2)_{\epsilon_8}\times(1/\nu_{4,m},\infty]\end{array} & \xrightarrow{\mathrm{Glue}_+^{(2)}} & \mathcal{M}_+^{\mathcal{E}_1^{(2)}\oplus\mathcal{E}_2^{(2)}}((\Sigma_\infty^{(2)},\vec{z}^{(2)});u_1^{(2)},u_2^{(2)})_{\epsilon_6,\nu_1}
\end{array}$$

PROOF. Theorem 8.16 implies that we can choose ϵ_8, $\nu_{4,m}$ so that the lower horizontal arrow $\mathrm{Glue}_+^{(2)}$ exists for given ϵ_6 and ν_1. Here we use the $W^2_{m+1,\delta}$ space to work out the gluing analysis. Note the gluing map itself is independent of the Sobolev exponent but the domain of its convergence depends on it.

Since $((\Sigma_\infty^{(1)}, \vec{z}^{(1)}), u_\infty^{(1)})$ is an element of

$$\mathcal{M}_+^{\mathcal{E}_1^{(2)} \oplus \mathcal{E}_2^{(2)}}((\Sigma_\infty^{(2)}, \vec{z}_\infty^{(2)}); u_1^{(2)}, u_2^{(2)})_{\epsilon_7/2, \nu_2/2}$$

by Condition 8.22, we use (8.43) and Theorem 8.16 (1) to show that we may take the constants ϵ_9, $\nu_{3,m}$ so small that the natural inclusion induces the right vertical arrow of (8.49) and

$$\mathrm{Im}\left(\mathcal{M}_+^{\mathcal{E}_1^{(1)}\oplus\mathcal{E}_2^{(1)}}((\Sigma_\infty^{(1)},\vec{z}^{(1)});u_1^{(1)},u_2^{(1)})_{\epsilon_9,\nu_{3,m}}\right)$$

$$\subset \mathrm{Im}\begin{pmatrix} \mathcal{M}_+^{\mathcal{E}_1^{(2)}}((\Sigma_1^{(2)},\vec{z}_1^{(2)});u_1^{(2)})_{\epsilon_8}\ {}_{\mathrm{ev}_0}\times_{\mathrm{ev}_0} \\ \mathcal{M}_+^{\mathcal{E}_2^{(2)}}((\Sigma_2^{(2)},\vec{z}_2^{(2)});u_2^{(2)};\beta_2)_{\epsilon_8}\times(1/\nu_{4,m},\infty] \end{pmatrix}$$

in $\mathcal{M}_+^{\mathcal{E}_1^{(2)}\oplus\mathcal{E}_2^{(2)}}((\Sigma_\infty^{(2)},\vec{z}^{(2)});u_1^{(2)},u_2^{(2)})_{\epsilon_6,\nu_1}$.

Then, using Theorem 8.16 (1) (the surjectivity and injectivity of the gluing map), we may take $\epsilon_{10,m}$ so small and $T_{m,(8.49)}$ so large that there exists a unique map $\mathfrak{F}$ so that Diagram (8.49) commutes. □

We take $m_0 > 10$ and fix it. In the next theorem we take $\epsilon \le \epsilon_{10,m_0}$ and consider the map $\mathfrak{F}$ as in (8.49) for $m = m_0$.

THEOREM 8.25. *There exists $T_{m_0,\langle\langle 8.25\rangle\rangle} > 0$ with the following property. We identify $(T(1),\infty]$ with $[0,1/T(1))$ under the map $T \mapsto s = 1/T$. Then the map $\mathfrak{F}$ in (8.49) is a smooth embedding on*

$$\mathcal{M}_+^{\mathcal{E}_1^{(1)}}((\Sigma_1^{(1)},\vec{z}_1^{(1)});u_1^{(1)})_{\epsilon}\ {}_{\mathrm{ev}_0}\times_{\mathrm{ev}_0} \mathcal{M}_+^{\mathcal{E}_2^{(1)}}((\Sigma_2^{(1)},\vec{z}_2^{(1)});u_2^{(1)})_{\epsilon}\times(T(1),\infty]$$

if $T(1) > T_{m_0,\langle\langle 8.25\rangle\rangle}$.

REMARK 8.26. We remark that the obstruction bundle data include the data of analytic family of coordinates at the 0-th marked point, which are used to define the obstruction bundle $\mathcal{E}_i^{(j)}(u')$. This way of defining the obstruction bundle is essential for the right vertical arrow of (8.49) to exist. In this way the obstruction bundle $\mathcal{E}_i^{(j)}(u')$ can be made to depend only on u', its source (marked bordered) curve, and the obstruction bundle data. In particular it is independent of the analytic family of coordinates $\varphi_i^{(j),\mathbb{R}}$, which we use to perform the inductive construction of the gluing map by the alternating method.

On the other hand the analytic families of coordinates at two chart centers are in general different from each other. Proposition 8.27 below is used to estimate the discrepancy between these two different choices.

PROOF. We observe that a neighborhood of $(\Sigma_\infty^{(j)},\vec{z}_\infty^{(j)}) \in \mathcal{M}_{g_1+g_2,k_1+k_2}$ is parameterized by the map (8.16), that is,

$$\mathrm{Gluesoc}^{(j),\mathbb{R}} : \mathcal{V}_1^{(j),\mathbb{R}} \times \mathcal{V}_2^{(j),\mathbb{R}} \times (T,\infty] \to \mathcal{CM}_{g_1+g_2,k_1+k_2}. \tag{8.50}$$

So we obtain a map

$$\mathcal{M}_+^{\mathcal{E}_1^{(j)}\oplus\mathcal{E}_2^{(j)}}((\Sigma_\infty^{(j)},\vec{z}_\infty^{(j)});u_1^{(j)},u_2^{(j)})_{\epsilon,\nu} \to \mathcal{V}_1^{(j),\mathbb{R}} \times \mathcal{V}_2^{(j),\mathbb{R}} \times (1/\nu,\infty] \tag{8.51}$$

by forgetting the map part of the stable map and compose it with the inverse of (8.50), for $j = 1,2$.

PROPOSITION 8.27. *There exist $\delta_3 > 0$, and a (strata-wise) smooth map*

$$\Phi : \mathcal{V}_1^{(1),\mathbb{R}} \times \mathcal{V}_2^{(1),\mathbb{R}} \times (1/\nu_{3,m},\infty] \to \mathcal{V}_1^{(2),\mathbb{R}} \times \mathcal{V}_2^{(2),\mathbb{R}} \times (1/\nu_1,\infty]$$

such that:

(1) *The next diagram commutes.*

(8.52)
$$\begin{array}{ccc}\mathcal{M}_+^{\mathcal{E}_1^{(1)}\oplus\mathcal{E}_2^{(1)}}((\Sigma_\infty^{(1)},\vec{z}_\infty^{(1)});u_1^{(1)},u_2^{(1)})_{\epsilon_9,\nu_{3,m}} & \longrightarrow & \mathcal{V}_1^{(1),\mathbb{R}}\times\mathcal{V}_2^{(1),\mathbb{R}}\times(1/\nu_{3,m},\infty] \\ \big\downarrow & & \big\downarrow \Phi \\ \mathcal{M}_+^{\mathcal{E}_1^{(2)}\oplus\mathcal{E}_2^{(2)}}((\Sigma_\infty^{(2)},\vec{z}_\infty^{(2)});u_1^{(2)},u_2^{(2)})_{\epsilon_6,\nu_1} & \longrightarrow & \mathcal{V}_1^{(2),\mathbb{R}}\times\mathcal{V}_2^{(2),\mathbb{R}}\times(1/\nu_1,\infty]\end{array}$$

Here the horizontal arrows are the forgetful maps (8.51), *the left vertical arrow is the right vertical arrow of Diagram* (8.49).

(2) *We put* $\Phi(\sigma;T)=(\sigma'(\sigma;T),T'(\sigma;T))$. *Then we have the following estimate.*

$$\begin{aligned}\left|\nabla_\sigma^n\frac{d^\ell}{dT^\ell}\sigma'(\sigma;T)\right| &\le C_{m,(8.53)}e^{-\delta_3 T},\\ \left|\nabla_\sigma^n\frac{d^\ell}{dT^\ell}(T'(\sigma;T)-T)\right| &\le C_{m,(8.53)}e^{-\delta_3 T}\end{aligned}\tag{8.53}$$

for $m-2\ge n$, $m-2\ge\ell>0$.

Note δ_3 in this proposition is the same constant as in Proposition 8.19.

PROOF. Existence of the map Φ satisfying (1) is an immediate consequence of the fact that during the inductive step of the construction of gluing map in Chapters 5 and 6, we never change the complex structure of the source. The estimate (8.53) will be proved in Section 8.4. □

We define

$$\begin{aligned}\mathrm{Resfor}^{(j)}:\mathcal{M}_+^{\mathcal{E}_1^{(j)}\oplus\mathcal{E}_2^{(j)}}&((\Sigma_\infty,\vec{z}_\infty);u_1^{(j)},u_2^{(j)})_{\epsilon,\nu}\\ &\to\mathcal{V}_1^{(j),\mathbb{R}}\times\mathcal{V}_2^{(j),\mathbb{R}}\times(1/\nu,\infty]\\ &\quad\times\prod_{i=1}^{2}\mathrm{Map}_{L^2_{m+1}}((K_i^{(j),S},K_i^{(j),S}\cap\partial\Sigma_i),(X,L))\end{aligned}\tag{8.54}$$

where the $\mathcal{V}_1^{(j),\mathbb{R}}\times\mathcal{V}_2^{(j),\mathbb{R}}\times(1/\nu,\infty]$ component of $\mathrm{Resfor}^{(j)}$ is the map (8.51) and its component of the factor in the third line is the restriction map and

$$K_i^{(j),S}=\Sigma_i^{(j)}\setminus\varphi_i^{(j),\mathbb{R}}(D^2_{\le 0}(1)).\tag{8.55}$$

We put

$$\begin{aligned}&\mathcal{U}^{(j)}_{\epsilon^{(j)},\nu^{(j)},T(2)}\\ &=(\mathrm{Glue}_+^{(j)})^{-1}\left(\mathcal{M}_+^{\mathcal{E}_1^{(j)}\oplus\mathcal{E}_2^{(j)}}((\Sigma_\infty^{(j)},\vec{z}_\infty^{(j)});u_1^{(j)},u_2^{(j)})_{\epsilon^{(j)},\nu^{(j)}}\right)\cap\pi_3^{-1}((T(2),\infty])\\ &\subset\mathcal{M}_+^{\mathcal{E}_1^{(j)}}((\Sigma_1^{(j)},\vec{z}_1^{(j)});u_1^{(j)})_{\epsilon'^{(j)}}\ {}_{\mathrm{ev}_0}\times_{\mathrm{ev}_0}\mathcal{M}_+^{\mathcal{E}_2^{(j)}}((\Sigma_2^{(j)},\vec{z}_2^{(j)});u_2^{(j)})_{\epsilon'^{(j)}}\times(T(2),\infty],\end{aligned}$$

where $(\epsilon^{(1)},\nu^{(1)},\epsilon'^{(1)})=(\epsilon_9,\nu_{3,m},\epsilon)$ with $\epsilon<\epsilon_{10,m}$, and $(\epsilon^{(2)},\nu^{(2)},\epsilon'^{(2)})=(\epsilon_6,\nu_1,\epsilon_8)$ and π_3 is the projection to the $(1/\nu,\infty]$ factor.

LEMMA 8.28. *Let* $m'\ge m+n, m\ge 2$. *We embed* $(T(2),\infty]$ *to* $[0,\nu^{(j)})$ *by* $T\mapsto s=1/T$. *Then*

$$\mathrm{Resfor}^{(j)}\circ\mathrm{Glue}_+^{(j)}:\mathcal{U}^{(j)}_{\epsilon^{(j)},\nu^{(j)},T(2)}\to\textit{(RHS of (8.54))}\tag{8.56}$$

is an embedding of C^n class if $T(2) > T_{(8.56),m'}$.

PROOF. The composition of $\mathrm{Resfor}^{(j)} \circ \mathrm{Glue}_+^{(j)}$ with the projection to the factor $\mathcal{V}_1^{(j),\mathbb{R}} \times \mathcal{V}_2^{(j),\mathbb{R}} \times (1/\nu^{(j)}, \infty]$ is smooth since it coincides with the projection to this factor. In other words the next diagram commutes.

$$
\begin{array}{ccc}
\mathcal{U}^{(j)}_{\epsilon^{(j)},\nu^{(j)},T(2)} & \longrightarrow & \mathcal{V}_1^{(j),\mathbb{R}} \times \mathcal{V}_2^{(j),\mathbb{R}} \times (T(2),\infty] \\
\Big\downarrow \mathrm{Resfor}^{(j)} \circ \mathrm{Glue}_+^{(j)} & & \Big\downarrow \\
\begin{array}{c}\mathcal{V}_1^{(j),\mathbb{R}} \times \mathcal{V}_2^{(j),\mathbb{R}} \times (T(2),\infty] \times \\ \times \prod_{i=1}^{2} \mathrm{Map}_{L^2_{m+1}}((K_i^{(j),S}, K_i^{(j),S} \cap \partial\Sigma_i),(X,L))\end{array} & \longrightarrow & \mathcal{V}_1^{(j),\mathbb{R}} \times \mathcal{V}_2^{(j),\mathbb{R}} \times (T(2),\infty],
\end{array}
\tag{8.57}
$$

where the second vertical arrow is the identity map and the horizontal arrows are projections. The commutativity of Diagram (8.57) follows from the fact that the construction of our map $\mathrm{Glue}_+^{(j)}$ does not change the complex structure of the source curve.

We define the map $\mathrm{Glue}_+^{(j)}$ using the $W^2_{m'+1,\delta}$ space. Its domain is given by $\infty \geq T > T_{m'}$. We consider the composition of $\mathrm{Resfor}^{(j)} \circ \mathrm{Glue}_+^{(j)}$ with the projection to the factor in the third line of the right hand side of (8.54). It becomes a map

$$
\mathcal{U}^{(j)}_{\epsilon^{(j)},\nu^{(j)},T(2)} \to \prod_{i=1}^{2} \mathrm{Map}_{L^2_{m+1}}((K_i^{(j),S}, K_i^{(j),S} \cap \partial\Sigma_i),(X,L)). \tag{8.58}
$$

Then by Theorem 8.17 we can show that this map $\mathrm{Resfor}^{(j)} \circ \mathrm{Glue}_+^{(j)}$ is of C^n class. Here we use our assumption $m' \geq m+n$ on the parts $T < \infty$ and $T = \infty$.

We next discuss differentiability at $T = \infty$. Here we use $s = 1/T$ instead of T for the coordinate of the $[0, \nu^{(j)})$ factor.

SUBLEMMA 8.29. *We have*

$$
\lim_{s_0 \to 0} \left\| \nabla^n_{\sigma,\rho} \frac{d^\ell}{ds^\ell} (8.58) \Big|_{s=s_0} \right\|_{L^2_{m+1}} = 0,
$$

for $m' \geq m+n$, $m \geq 2$, $m' - m \geq \ell > 0$.

PROOF. Note

$$
\frac{d^\ell}{ds^\ell} = \frac{1}{s^{2\ell}} \mathcal{Q}\left(s, \frac{d}{dT}\right) \frac{d}{dT},
$$

where $\mathcal{Q}(x,\xi) = \sum_{i=0}^{\ell-1} Q_i(x)\xi^i$ and Q_i are polynomials. Therefore by the exponential decay provided in Theorem 8.17 we derive

$$
\left\| \nabla^n_{\sigma,\rho} \frac{d^\ell}{ds^\ell} (8.58) \Big|_{s=s_0} \right\|_{L^2_{m+1}} \leq C_{m,(8.59)} s_0^{-2\ell} e^{-\delta_2/s_0}. \tag{8.59}
$$

The sublemma follows. □

Thus the map (8.56) is of C^n class.

To prove that it is an *embedding*, we use Diagram (8.57). In view of its commutativity, it suffices to show that $\mathrm{pr} \circ \mathrm{Resfor}^{(j)} \circ \mathrm{Glue}_+^{(j)}$ is a smooth embedding when we restrict it to the fiber of an arbitrary point (σ_1, σ_2, T) in $\mathcal{V}_1^{(j),\mathbb{R}} \times \mathcal{V}_2^{(j),\mathbb{R}} \times (1/\nu^{(j)}, \infty]$. Here pr is the projection to the factor in the third line of (8.54).

At $T = \infty$ (or $s = 0$), the map $\mathrm{pr} \circ \mathrm{Resfor}^{(j)} \circ \mathrm{Glue}_+^{(j)}$ is actually the restriction map, since, there, $\mathrm{Glue}_+^{(j)}$ is obtained by identifying two stable maps at the 0-th marked points. Therefore it is an embedding by unique continuation of pseudo-holomorphic curve.

On the other hand, (8.59) implies that the restriction of the map $\mathrm{Resfor}^{(j)} \circ \mathrm{Glue}_+^{(j)}$ to the fiber of (σ_1, σ_2, T) converges to its restriction to $(\sigma_1, \sigma_2, \infty)$ in the C^1 sense as T goes to infinity. Therefore we obtain the required properties for sufficiently large $T(2)$. □

Now we are in the position to complete the proof of Theorem 8.25.

Let $m_0 \geq 10$ and $m' \geq m_1 + n$, $m_1 \geq m_0 + n$. We consider the composition of the following three maps. The first map to be composed is

$$\begin{aligned}\mathrm{Glue}_+^{(1)} : \mathcal{M}_+^{\mathcal{E}_1^{(1)}}((\Sigma_1^{(1)}, \vec{z}_1^{(1)}); u_1^{(1)})_{\epsilon\ \mathrm{ev}_0} \times_{\mathrm{ev}_0} \mathcal{M}_+^{\mathcal{E}_2^{(1)}}((\Sigma_2^{(1)}, \vec{z}_2^{(1)}); u_2^{(1)})_{\epsilon} \times (T(3), \infty] \\ \to \mathcal{M}_+^{\mathcal{E}_1^{(1)} \oplus \mathcal{E}_2^{(1)}}((\Sigma_\infty^{(1)}, \vec{z}_\infty^{(1)}); u_1^{(1)}, u_2^{(1)})_{\epsilon_9, \nu_{3,m'}},\end{aligned}$$

where $\epsilon < \epsilon_{10,m'}$. The second map is the inclusion

$$\mathcal{M}_+^{\mathcal{E}_1^{(1)} \oplus \mathcal{E}_2^{(1)}}((\Sigma_\infty^{(1)}, \vec{z}_\infty^{(1)}); u_1^{(1)}, u_2^{(1)})_{\epsilon_9, \nu_{3,m'}} \to \mathcal{M}_+^{\mathcal{E}_1^{(2)} \oplus \mathcal{E}_2^{(2)}}((\Sigma_\infty^{(2)}, \vec{z}_\infty^{(2)}); u_1^{(2)}, u_2^{(2)})_{\epsilon_6, \nu_1}.$$

The third map is

$$\begin{aligned}\mathrm{pr} \circ \mathrm{Resfor}^{(2)} : \mathcal{M}_+^{\mathcal{E}_1^{(2)} \oplus \mathcal{E}_2^{(2)}}&((\Sigma_\infty^{(2)}, \vec{z}_\infty^{(2)}); u_1^{(2)}, u_2^{(2)})_{\epsilon_6, \nu_1} \\ &\to \mathcal{V}_1^{(2),\mathbb{R}} \times \mathcal{V}_2^{(2),\mathbb{R}} \times (1/\nu_1, \infty] \\ &\qquad \times \prod_{i=1}^{2} \mathrm{Map}_{L^2_{m_0+1}}((K_i^{(2)}, K_i^{(2),S} \cap \partial \Sigma_i), (X, L)) \\ &\to \prod_{i=1}^{2} \mathrm{Map}_{L^2_{m_0+1}}((K_i^{(2)}, K_i^{(2),S} \cap \partial \Sigma_i), (X, L)).\end{aligned}$$

LEMMA 8.30. *The composition of the above three maps is of C^n class. Its restriction to the fiber of the map*

$$\begin{aligned}&\mathcal{M}_+^{\mathcal{E}_1^{(1)}}((\Sigma_1^{(1)}, \vec{z}_1^{(1)}); u_1^{(1)})_{\epsilon\ \mathrm{ev}_0} \times_{\mathrm{ev}_0} \mathcal{M}_+^{\mathcal{E}_2^{(1)}}((\Sigma_2^{(1)}, \vec{z}_2^{(1)}); u_2^{(1)})_{\epsilon} \times (T(3), \infty] \\ &\to \mathcal{V}_1^{(1),\mathbb{R}} \times \mathcal{V}_2^{(1),\mathbb{R}} \times (1/\nu_1, \infty],\end{aligned}$$

is an embedding of C^n-class for all $T(3) \geq T_{m', \langle\langle 8.30 \rangle\rangle}$.

To prove the lemma we extend the map ψ_{σ_i} in (8.47) including the T parameter. Let $(\sigma, T) = ((\sigma_1, \sigma_2), T) \in \mathcal{V}_1^{(1),\mathbb{R}} \times \mathcal{V}_2^{(1),\mathbb{R}} \times (1/\nu_1, \infty]$. We obtain $\Phi(\sigma, T) = (\sigma'(\sigma; T), T'(\sigma; T))$ as in Proposition 8.27. We have

$$\Sigma_1^{(1),\sigma_1} \#_T \Sigma_2^{(1),\sigma_2} \cong \Sigma_1^{(2),\sigma_1'} \#_{T'} \Sigma_2^{(2),\sigma_2'}.$$

Here we use the gluing datum $\Xi^{(1)}$ (resp. $\Xi^{(2)}$) to define the left hand side (resp. right hand side). We consider the composition

$$\widehat{\Psi}_{i,\sigma,T} : K_i^{(2)} \subset \Sigma_1^{(2),\sigma_1'} \#_{T'} \Sigma_2^{(2),\sigma_2'} \cong \Sigma_1^{(1),\sigma_1} \#_T \Sigma_2^{(1),\sigma_2}$$

where the first map is the canonical embedding. By definition $\widehat{\Psi}_{i,\sigma,\infty}$ coincides with the composition of ψ_{σ_i} and the canonical embedding : $\Sigma_i^{(2)} \to \Sigma_1^{(2),\sigma_1'} \#_\infty \Sigma_2^{(2),\sigma_2'}$. Therefore by Condition 8.23 we may assume that there exists a map

$$\Psi_{i,\sigma,T} : K_i^{(2)} \to K_i^{(1)}$$

such that $\widehat{\Psi}_{i,\sigma,T}$ is the composition of $\Psi_{i,\sigma,T}$ and the canonical embedding $K_i^{(1)} \to \Sigma_1^{(1),\sigma_1} \#_T \Sigma_2^{(1),\sigma_2}$.

This map is the special case of the map (8.73) and the estimate (8.76) in Proposition 8.35 holds. Namely we have

$$\left\| \nabla_\sigma^n \frac{d^\ell}{dT^\ell} \Psi_{i,\sigma,T} \right\|_{C^\ell(K_i^{(a)}, K_i^{(b)})} \le C_{n,\ell,(8.60)} e^{-\delta_4 T} \tag{8.60}$$

for $\ell > 0$.

PROOF OF LEMMA 8.30. The composition of three maps is nothing but the composition of the following two maps (8.61) and (8.62). The first map is the map

$$\begin{aligned} &\mathrm{Resfor}^{(1)} \circ \mathrm{Glue}_+^{(1)} : \\ &\mathcal{M}_+^{\mathcal{E}_1^{(1)}}((\Sigma_1^{(1)}, \vec{z}_1^{(1)}); u_1^{(1)})_{\epsilon \mathrm{ev}_0} \times_{\mathrm{ev}_0} \mathcal{M}_+^{\mathcal{E}_2^{(1)}}((\Sigma_2^{(1)}, \vec{z}_2^{(1)}); u_2^{(1)})_\epsilon \times (T(3), \infty] \\ &\to \mathcal{V}_1^{(1),\mathbb{R}} \times \mathcal{V}_2^{(1),\mathbb{R}} \times (1/\nu_1, \infty] \times \prod_{i=1}^2 \mathrm{Map}_{L^2_{m_1+1}}((K_i^{(1)}, K_i^{(1)} \cap \partial \Sigma_i), (X, L)). \end{aligned} \tag{8.61}$$

The second map is

$$\begin{aligned} &\mathcal{V}_1^{(1),\mathbb{R}} \times \mathcal{V}_2^{(1),\mathbb{R}} \times (1/\nu_1, \infty] \times \prod_{i=1}^2 \mathrm{Map}_{L^2_{m_1+1}}((K_i^{(1)}, K_i^{(1)} \cap \partial \Sigma_i), (X, L)) \\ &\qquad \to \prod_{i=1}^2 \mathrm{Map}_{L^2_{m_0+1}}((K_i^{(2)}, K_i^{(2)} \cap \partial \Sigma_i), (X, L)) \end{aligned} \tag{8.62}$$

which is defined by

$$((\sigma_1, \sigma_2, T), (v_1, v_2)) \mapsto (v_1 \circ \Psi_{1,\sigma,T}, v_2 \circ \Psi_{2,\sigma,T}).$$

Lemma 8.28 in the case $j = 1$ implies that the first map (8.61) is of C^n class. We can use (8.60) and $m_1 \ge m_0 + n$ to show the second map (8.62) is also of C^n class. Thus the composition of the three maps is of C^n class.

The fact that it is an embedding on the fiber follows from Lemma 8.28. □

Using Proposition 8.27 (especially the estimate (8.53)) and Lemma 8.28 applied to $\mathrm{Resfor}^{(2)} \circ \mathrm{Glue}_+^{(2)}$, Lemma 8.30 imply $\mathfrak{F}$ in (8.49) is of C^n class.

We finally prove that $\mathfrak{F}$ is of C^∞ class. This part of the proof is a bit sketchy. We refer the reader to [**FOOO2**, Section 26] or [**FOOO3**, Section 12] for detail. We fix m_0. For $j = 1, 2$, we take a C^∞ structure of

$$\mathcal{M}_+^{\mathcal{E}_1^{(j)}}((\Sigma_1^{(i)}, \vec{z}_1^{(j)}); u_1^{(i)})_{\epsilon\ \mathrm{ev}_0} \times_{\mathrm{ev}_0} \mathcal{M}_+^{\mathcal{E}_2^{(j)}}((\Sigma_2^{(i)}, \vec{z}_2^{(j)}); u_2^{(i)})_\epsilon \times (T(3), \infty] \tag{8.63}$$

as follows. (Here $\epsilon \le \epsilon_{10,m_0}$.) We consider the composition

$$\begin{aligned}\mathrm{Resfor}^{(j)} \circ \mathrm{Glue}_+^{(j)} : (8.63) \to & \mathcal{V}_1^{(j),\mathbb{R}} \times \mathcal{V}_2^{(j),\mathbb{R}} \times (1/\nu_1, \infty] \\ & \times \prod_{i=1}^{2} \mathrm{Map}_{L^2_{m_0+1}}((K_i^{(j),S}, K_i^{(j),S} \cap \partial\Sigma_i), (X, L)).\end{aligned}$$

We claim that the image of this map is a submanifold of C^∞ class of the codomain. This is an immediate consequence of elliptic regularity on the part $T \ne \infty$.

At $T = \infty$ we use Lemma 8.28 for various $m' > m = m_0$ to prove the image of $\mathrm{Resfor}^{(j)} \circ \mathrm{Glue}_+^{(j)}$ is smooth there.

Now we *define* the C^∞ structure of (8.63) so that $\mathrm{Resfor}^{(j)} \circ \mathrm{Glue}_+^{(j)}$ is a C^∞ embedding. Note that the smooth structure on $(1/\nu_1, \infty]$ is given so that $T \in (1/\nu_1, \infty] \mapsto 1/T \in [0, \nu_1)$ is a diffeomorphism.

Then using this new C^∞ structure the composition in Lemma 8.30 for various m', m_1 we can prove that $\mathfrak{F}$ is an embedding of C^∞ class as follows. On the part $T \ne \infty$ it is a consequence of elliptic regularity. At $T = \infty$ it suffices to show that it is of C^n class for any n at an arbitrary point with $T = \infty$. We take $m_1 = m_0 + n$ and $m' = m_1 + n$ for various n, and apply Lemma 8.30 to show that $\mathfrak{F}$ is of C^n class there. The proof of Theorem 8.25 is now complete. □

In a similar way we can prove the smoothness of the Kuranishi map as follows. Hereafter we write

$$V^{(j)} = \mathcal{M}_+^{\mathcal{E}_1^{(j)}}((\Sigma_1^{(j)}, \vec{z}_1^{(j)}); u_1^{(j)})_{\epsilon^{(j)},\nu^{(j)}} \times_L \mathcal{M}_+^{\mathcal{E}_2^{(j)}}((\Sigma_2^{(j)}, \vec{z}_2^{(j)}); u_2^{(j)})_{\epsilon^{(j)},\nu^{(j)}}, \tag{8.64}$$

where $(\epsilon^{(1)}, \nu^{(1)}, \epsilon'^{(1)}) = (\epsilon_9, \nu_{3,m}, \epsilon)$ with $\epsilon < \epsilon_{10}$, and $(\epsilon^{(2)}, \nu^{(2)}, \epsilon'^{(2)}) = (\epsilon_6, \nu_1, \epsilon_8)$.

We consider the space $\mathcal{M}_+^{\mathcal{E}_1^{(j)} \oplus \mathcal{E}_2^{(j)}}((\Sigma_\infty^{(1)}, \vec{z}_\infty^{(j)}); u_1^{(j)}, u_2^{(j)})_{\epsilon'^{(j)},\nu^{(j)}}$. We regard the image of

$$\mathrm{Glue}_+^{(j)} : V^{(j)} \times [0, \nu^{(j)}) \to \mathcal{M}_+^{\mathcal{E}_1^{(j)} \oplus \mathcal{E}_2^{(j)}}((\Sigma_\infty^{(j)}, \vec{z}_\infty^{(j)}); u_1^{(j)}, u_2^{(j)})_{\epsilon'^{(j)},\nu^{(j)}}$$

as a smooth manifold so that $\mathrm{Glue}_+^{(j)}$ is a diffeomorphism.

Let $\mathfrak{x} = ((\Sigma'_T, \vec{z}), u') \in \mathcal{M}_+^{\mathcal{E}_1^{(j)} \oplus \mathcal{E}_2^{(j)}}((\Sigma_\infty^{(j)}, \vec{z}_\infty^{(j)}); u_1^{(j)}, u_2^{(j)})_{\epsilon'^{(j)},\nu^{(j)}}$. The definition of the moduli space implies:

$$\overline{\partial} u' \equiv 0, \qquad \mathrm{mod}\ \mathcal{E}_1^{(j)}(u') \oplus \mathcal{E}_2^{(j)}(u'). \tag{8.65}$$

We write the left hand side as $\widetilde{\mathfrak{s}}^{(j)}(\mathfrak{x})$. By the inverse of the map $I_{u',i}$ in (8.19) we obtain

$$\mathfrak{s}^{(j)}(\mathfrak{x}) = (I_{u',1} \oplus I_{u',2})^{-1}(\widetilde{\mathfrak{s}}^{(j)}(\mathfrak{x})) \in \mathcal{E}_1^{(j)} \oplus \mathcal{E}_2^{(j)}.$$

Note $\mathcal{E}_1^{(j)} \oplus \mathcal{E}_2^{(j)}$ is independent of $\mathfrak{x}$.

We thus obtain a map

$$\mathfrak{s}^{(j)} \circ \mathrm{Glue}_+^{(j)} : V^{(j)} \times [0, \nu^{(j)}) \to \mathcal{E}_1^{(j)} \oplus \mathcal{E}_2^{(j)}. \tag{8.66}$$

This is by definition the *Kuranishi map*.

PROPOSITION 8.31. *The map $\mathfrak{s}^{(j)}$ in (8.66) is smooth.*

PROOF. The smoothness at each of the stratum ($T = \infty$ and $T \ne \infty$) follows from elliptic regularity. Note the supports of elements of $\mathcal{E}_i^{(j)}(u')$ are in the image of $K_i^{\mathfrak{ob}}$. Therefore we can use (8.30) to show that all the derivatives of $\mathfrak{s}_i$ including at least one $s = 1/T$ derivatives vanishes at $s = 0$ in the same way as in the proof of Sublemma 8.29. Furthermore (8.30) implies that the restriction of $\mathfrak{s}^{(j)}$ to $s = s_0$ converges to its restriction to $s = 0$ in the C^n sense for any n as s_0 goes to 0. The proof of Proposition 8.31 is complete. □

We are finally ready to provide a construction of the Kuranishi neighborhoods, the associated Kuranishi maps and parametrization maps, and to wrap-up the proof of smoothness of coordinate change maps. We put $\hat{V}^{(j)} = V^{(j)} \times [0, \nu^{(j)})$ and $\widehat{\mathcal{E}}^{(j)} = \mathcal{E}_1^{(j)} \oplus \mathcal{E}_2^{(j)}$. Then we can find $\psi^{(j)}$ so that the quintuple $(\hat{V}^{(j)}, \widehat{\mathcal{E}}^{(j)}, \{1\}, \psi^{(j)}, \mathfrak{s}^{(j)})$ becomes a Kuranishi neighborhood of the moduli space of stable bordered maps $\mathcal{M}_{g_1+g_2,k_1+k_2}(X, L; \beta_1 + \beta_2)$ in the sense of [**FOOO1**, Definition A1.1].

In fact, the moduli space $\mathcal{M}_{g_1+g_2,k_1+k_2}(X, L; \beta_1 + \beta_2)$ (See Definition 8.8) is locally identified with the zero set of $\mathfrak{s}^{(j)}$. This fact follows from the 'injectivity' and 'surjectivity' we proved in Chapter 7. Therefore we obtain our parametrization map

$$\psi^{(j)} : (\mathfrak{s}^{(j)})^{-1}(0) \to \mathcal{M}_{g_1+g_2,k_1+k_2}(X, L; \beta_1 + \beta_2).$$

Moreover the group of automorphisms of the objects in our neighborhood in the moduli space $\mathcal{M}_{g_1+g_2,k_1+k_2}(X, L; \beta_1 + \beta_2)$ is trivial. Hence we can put $\Gamma = \{1\}$.

Now we prove:

THEOREM 8.32. *There exists a smooth coordinate change*

$$(\phi_{21}, \hat{\phi}_{21}, \mathrm{id}) : (\hat{V}^{(1)}, \widehat{\mathcal{E}}^{(1)}, \{1\}, \psi^{(1)}, \mathfrak{s}^{(1)}) \to (\hat{V}^{(2)}, \widehat{\mathcal{E}}^{(2)}, \{1\}, \psi^{(2)}, \mathfrak{s}^{(2)})$$

of Kuranishi neighborhoods in the sense of [**FOOO1**, Definition A.1.3].

PROOF. The map $\phi_{21} : \hat{V}^{(1)} \to \hat{V}^{(2)}$ is nothing but the embedding $\mathfrak{F}$ in Diagram (8.49). This map is a smooth embedding by Theorem 8.25.

The bundle map $\hat{\phi}_{21} : \mathcal{E}^{(1)} \to \mathcal{E}^{(2)}$ is obtained from (8.43), that is, $\mathcal{E}_1^{(1)} \oplus \mathcal{E}_2^{(1)} \subset \mathcal{E}_1^{(2)} \oplus \mathcal{E}_2^{(2)}$, as follows.

If $\mathfrak{x} \in \hat{V}^{(1)}$ and u' is the map part of

$$\mathrm{Glue}_+^{(1)}(\mathfrak{x}) \in \mathcal{M}_+^{\mathcal{E}_1^{(1)} \oplus \mathcal{E}_2^{(1)}}((\Sigma_\infty^{(1)}, \vec{z}_\infty^{(1)}); u_1^{(1)}, u_2^{(1)})_{\epsilon,\nu}$$

then

$$\mathcal{E}_1(\mathfrak{x}) = \mathcal{E}_1^{(1)}(u') \oplus \mathcal{E}_2^{(1)}(u') \subset \mathcal{E}_1^{(2)}(u') \oplus \mathcal{E}_2^{(2)}(u').$$

Note since u' is also the map part of $\mathrm{Glue}_+^{(2)}(\phi_{21}(\mathfrak{x}))$, we have

$$\mathcal{E}_2(\mathrm{Glue}_+^{(2)}(\phi_{21}(\mathfrak{x}))) = \mathcal{E}_1^{(2)}(u') \oplus \mathcal{E}_2^{(2)}(u').$$

The fiber of $\hat{\phi}_{21}$ at $\mathfrak{x}$ is the obvious inclusion $\mathcal{E}_i^{(1)}(\mathfrak{x}) \subset \mathcal{E}_i^{(2)}(\mathrm{Glue}_+^{(2)}(\phi_{21}(\mathfrak{x})))$.

The smoothness of $\hat{\phi}_{21}$ is proved in the same way as in Theorem 8.25 and Proposition 8.31 using the fact that the supports of elements of $\mathcal{E}_i^{\mathfrak{ob}}$ are in $K_i^{\mathfrak{ob}}$.

The group homomorphism id : $\{1\} \to \{1\}$ is the identity map. Various commutativities of compositions of maps required in [**FOOO1**, Definition A.1.3] are trivial to check in our case. □

REMARK 8.33. In this paper we assume that $(\Sigma_i, \vec{z}_i)$ is stable. For the unstable case we need to add marked points $\vec{z}_i^{+}$ and use the slices of codimension 2 submanifolds of X to reduce the construction of the Kuranishi chart to the case when the source is stable. (We use the slices that are transversal to the map $u_i : \Sigma_i \to X$ at

$\vec{z}_i^{+}$, to cut down the moduli space to one of correct dimension. See [**FOn**, Appendix] and [**FOOO2**, Part IV].)

We also need to show that the change of the choices of extra marked points and codimension 2 submanifolds, induces a *smooth* coordinate change. Once the estimates (8.30) and (8.53) are established the rest of the proof of this statement is rather geometric than analytic. So we do not discuss this point in this paper, whose focus lies in the analytic part of the story. See [**FOOO1**, page 772], [**FOOO2**, Part IV] and [**FOOO3**, Sections 10,12] for the argument on this point.

8.4. Comparison between two choices of analytic families of coordinates

In this section we prove Propositions 8.19 and 8.27 (2).

The main part of the proof is Proposition 8.35 below. We first need to set up notations to state it. Let $(\Sigma_i^{(j)}, \vec{z}_i^{(j)}) \in \mathcal{M}_{g_i,k_i+1}$, $i = 1,2$, $j = a,b$. We choose gluing data $\Xi^{(j)} = (\Xi_1^{(j)}, \Xi_2^{(j)})$ centered at $([\Sigma_1^{(j)}, \vec{z}_1^{(j)}], [\Sigma_2^{(j)}, \vec{z}_2^{(j)}])$ for $j = a$ and $j = b$. They induce the source gluing maps:

$$\begin{aligned} \mathrm{Gluesoc}^{\mathbb{R}}_{\Xi^{(a)}} &: \mathcal{V}_1^{(a,+),\mathbb{R}} \times \mathcal{V}_2^{(a,+),\mathbb{R}} \times (T_0,\infty] \to \mathcal{CM}_{g_1+g_2,k_1+k_2}, \\ \mathrm{Gluesoc}^{\mathbb{R}}_{\Xi^{(b)}} &: \mathcal{V}_1^{(b),\mathbb{R}} \times \mathcal{V}_2^{(b),\mathbb{R}} \times (T_0,\infty] \to \mathcal{CM}_{g_1+g_2,k_1+k_2} \end{aligned} \tag{8.67}$$

by Definition-Lemma 8.7. Here $\mathcal{V}_i^{(a,+),\mathbb{R}}$, $\mathcal{V}_i^{(b),\mathbb{R}}$ are open neighborhoods of $[\Sigma_i^{(a)}, \vec{z}_i^{(a)}]$, $[\Sigma_i^{(b)}, \vec{z}_i^{(b)}]$ in $\mathcal{M}_{g_i,k_i+1}$, which are parts of $\Xi_i^{(a)}$, $\Xi_i^{(b)}$, respectively.

REMARK 8.34. In (8.16) the third factor of the domain is $[0,\epsilon)$. It is related to $(T_0,\infty]$ by $T \mapsto r = e^{-10\pi T}$. See (8.18).

We put:

$$\mathcal{V}_i^{(a),\mathbb{R}} = \mathcal{V}_i^{(a,+),\mathbb{R}} \cap \mathcal{V}_i^{(b),\mathbb{R}}, \tag{8.68}$$

and regard it as an open subset of $\mathcal{V}_i^{(a,+),\mathbb{R}}$. Then we may choose T_j and a map Φ so that the next diagram commutes.

$$\begin{array}{ccc} \mathcal{V}_1^{(a),\mathbb{R}} \times \mathcal{V}_2^{(a),\mathbb{R}} \times (T_a,\infty] & \xrightarrow{\mathrm{Gluesoc}^{\mathbb{R}}_{\Xi^{(a)}}} & \mathcal{CM}_{g_1+g_2,k_1+k_2} \\ \Phi \big\downarrow & & \big\downarrow \mathrm{id} \\ \mathcal{V}_1^{(b),\mathbb{R}} \times \mathcal{V}_2^{(b),\mathbb{R}} \times (T_b,\infty] & \xrightarrow{\mathrm{Gluesoc}^{\mathbb{R}}_{\Xi^{(b)}}} & \mathcal{CM}_{g_1+g_2,k_1+k_2} \end{array} \tag{8.69}$$

Let $K_i^{(j)} = \Sigma_i^{(j)} \setminus \mathrm{Im}(\varphi_i^{(j),\mathbb{R}})$. For $\sigma \in \mathcal{V}_i^{(a),\mathbb{R}}$ the canonical embedding associated to $\Xi^{(a)}$ induces

$$\mathfrak{I}^{\sigma}_{\Xi^{(a)},i} : K_i^{(a)} \to \Sigma_1^{(a),\sigma} \#_T \Sigma_2^{(a),\sigma}. \tag{8.70}$$

On the other hand, the canonical embedding associated to $\Xi^{(b)}$ induces

$$\mathfrak{I}^{\sigma'}_{\Xi^{(b)},i} : K_i^{(b)} \subset \Sigma_1^{(b),\sigma'} \#_T \Sigma_2^{(b),\sigma'}. \tag{8.71}$$

Let $(\sigma', T') = \Phi(\sigma, T)$, that is

$$\Sigma_1^{(a),\sigma} \#_T \Sigma_2^{(a),\sigma} \cong \Sigma_1^{(b),\sigma'} \#_{T'} \Sigma_2^{(b),\sigma'}.$$

Next we assume the inclusion

$$\mathfrak{I}^{\sigma}_{\Xi^{(a)},i}(K_i^{(a)}) \subset \mathfrak{I}^{\sigma'}_{\Xi^{(b)},i}(K_i^{(b)}). \tag{8.72}$$

We then obtain a smooth embedding

$$\Psi_{i,\sigma,T} = (\mathfrak{I}^{\sigma'}_{\Xi^{(b)},i})^{-1} \circ \mathfrak{I}^{\sigma}_{\Xi^{(a)},i} : K_i^{(a)} \to K_i^{(b)}. \tag{8.73}$$

This defines a (σ, T)-family of maps which we regard as the map

$$\Psi_i : \mathcal{V}_1^{(a),\mathbb{R}} \times \mathcal{V}_2^{(a),\mathbb{R}} \times (T_a, \infty] \times K_i^{(a)} \to K_i^{(b)}. \tag{8.74}$$

PROPOSITION 8.35. *There exists $\delta_4 > 0$ with the following properties.*

(1) *Consider the map Φ defined by $\Phi(\sigma;T) = (\sigma'(\sigma;T), T'(\sigma;T))$. Then we have the following estimate.*

$$\begin{aligned} \left|\nabla^n_\sigma \frac{d^\ell}{dT^\ell}\sigma'(\sigma;T)\right| &\le C_{n,\ell,(8.75)} e^{-\delta_4 T}, \\ \left|\nabla^n_\sigma \frac{d^\ell}{dT^\ell}(T'(\sigma;T) - T)\right| &\le C_{n,\ell,(8.75)} e^{-\delta_4 T} \end{aligned} \tag{8.75}$$

for $\ell > 0$.

(2) *We assume (8.72). Let Ψ_i be the map given in (8.74). Then*

$$\left\|\nabla^n_\sigma \frac{d^\ell}{dT^\ell}\Psi_i\right\|_{C^\ell(K_i^{(a)},K_i^{(b)})} \le C_{n,\ell,(8.76)} e^{-\delta_4 T} \tag{8.76}$$

for $\ell > 0$.

PROOF. We first prove (8.75). By taking the double of (8.69) we obtain the next diagram.

$$\begin{array}{ccc} \mathcal{V}_1^{(a)} \times \mathcal{V}_2^{(a)} \times D^2(\epsilon) & \xrightarrow{\text{Gluesoc}^{\mathbb{C}}_{\Xi^{(a)}}} & \mathcal{CM}^{\text{cl}}_{g_1+g_2,k_1+k_2} \\ \Phi^{\mathbb{C}}\downarrow & & \text{id}\downarrow \\ \mathcal{V}_1^{(b)} \times \mathcal{V}_2^{(b)} \times D^2(\epsilon) & \xrightarrow{\text{Gluesoc}^{\mathbb{C}}_{\Xi^{(b)}}} & \mathcal{CM}_{g_1+g_2,k_1+k_2} \end{array} \tag{8.77}$$

Since $\varphi_i^{(j),\mathbb{R}}$ are analytic coordinates $\text{Gluesoc}^{\mathbb{C}}_{\Xi^{(a)}}$ and $\text{Gluesoc}^{\mathbb{C}}_{\Xi^{(b)}}$ are holomorphic. Therefore $\Phi^{\mathbb{C}}$ is also holomorphic. We remark that

$$\Phi^{\mathbb{C}}(\mathcal{V}_1^{(a)} \times \mathcal{V}_2^{(a)} \times \{0\}) \subseteq (\mathcal{V}_1^{(b)} \times \mathcal{V}_2^{(b)} \times \{0\}).$$

Therefore there exist holomorphic maps $\mathcal{F} : \mathcal{V}_1^{(a)} \times \mathcal{V}_2^{(a)} \to \mathcal{V}_1^{(b)} \times \mathcal{V}_2^{(b)}$ and $\mathcal{G} = (\mathcal{G}_1, \mathcal{G}_2) : \mathcal{V}_1^{(a)} \times \mathcal{V}_2^{(a)} \times D^2(1) \to \mathbb{C}^{\dim \mathcal{V}_1^{(b)} + \dim \mathcal{V}_2^{(b)}} \times \mathbb{C}$ such that

$$\Phi^{\mathbb{C}}(\sigma_1, \sigma_2, \mathfrak{r}) = (\mathcal{F}(\sigma_1, \sigma_2), 0) + \mathfrak{r}\mathcal{G}(\sigma_1, \sigma_2, \mathfrak{r})$$

Moreover $\mathcal{G}_2(\sigma_1, \sigma_2, 0) \ne 0$. Here we regard $\mathcal{V}_i^{(j)}$ as an open set of $\mathbb{C}^{\dim \mathcal{V}_i^{(j)}}$.

Note Φ is a restriction of $\Phi^{\mathbb{C}}$ where the T coordinate is identified with a part of the standard coordinate $\mathfrak{r}$ of $D^2(\epsilon)$ by the map $T \mapsto \mathfrak{r} = -e^{-10\pi T} \in D^2(\epsilon)$. (See Diagram (8.17) and Formula (8.18).) Therefore

$$\begin{aligned} \sigma'(\sigma, T) &= \mathcal{F}(\sigma) + e^{-10\pi T}\mathcal{G}_1(\sigma, -e^{-10\pi T}), \\ T'(\sigma, T) &= T - \log \mathcal{G}_2(\sigma, -e^{-10\pi T})/10\pi. \end{aligned}$$

Here $\sigma = (\sigma_1, \sigma_2)$. Now, the estimate (8.75) follows from the holomorphicity of $\mathcal{F}$ and $\mathcal{G}_1, \mathcal{G}_2$.

We next prove (2). By taking the double of (8.74) we obtain a map:

$$\Psi_i^{\mathbb{C}} : \mathcal{V}_1^{(a)} \times \mathcal{V}_2^{(a)} \times D^2(\epsilon) \times K_i^{(a)} \to K_i^{(b)}. \tag{8.78}$$

Using the trivialization of the universal bundle which is a part of the data $\Xi_i^{(j)}$ ($j = a, b$) we identify

$$\mathcal{V}_i^{(j)} \times K_i^{(j)} \subset \mathcal{C}(\mathcal{V}_i^{(j)}).$$

We use this embedding to define a complex structure of $\mathcal{V}_1^{(j)} \times \mathcal{V}_2^{(j)} \times K_i^{(j)}$.

We consider the map

$$\Phi^{\mathbb{C}} \times \Psi_i^{\mathbb{C}} : \mathcal{V}_1^{(a)} \times \mathcal{V}_2^{(a)} \times D^2(\epsilon) \times K_i^{(a)} \to \mathcal{V}_1^{(b)} \times \mathcal{V}_2^{(b)} \times D^2(\epsilon) \times K_i^{(b)}.$$

The map $\Phi^{\mathbb{C}} \times \Psi_i^{\mathbb{C}}$ is holomorphic with respect to the above complex structures because the left (resp. right) hand side is an open set of the universal family $\mathcal{C}(\mathcal{V}_1^{(a)} \times \mathcal{V}_2^{(a)} \times D^2(\epsilon))$ (resp. $\mathcal{C}(\mathcal{V}_1^{(b)} \times \mathcal{V}_2^{(b)} \times D^2(\epsilon))$). Note neither the left nor the right hand side is the direct product as a complex manifold with respect to this complex structure.

On the other hand, the projection

$$\mathrm{pr} : \mathcal{V}_1^{(b)} \times \mathcal{V}_2^{(b)} \times D^2(\epsilon) \times K_i^{(b)} \to K_i^{(b)}$$

is certainly of C^∞ class. Now we have

$$\Psi_i(\sigma, T, z) = \mathrm{pr}(\Phi^{\mathbb{C}}(\sigma, -e^{-10\pi T}), \Psi_i^{\mathbb{C}}(\sigma, -e^{-10\pi T}, z)). \tag{8.79}$$

Using this holomorphicity of $(\Phi^{\mathbb{C}}, \Psi_i^{\mathbb{C}})$ and smoothness of pr we obtain the estimate (8.76). □

PROOF OF PROPOSITION 8.27. Statement (1) is already proved. Replacing a, b by $1, 2$ we apply Proposition 8.35 (1).[5] Note we use Proposition 8.35 (1) only here. So (8.72) is not necessary. Then (8.53) follows from (8.75). □

PROOF OF PROPOSITION 8.19. The map $I_{u_i^{\sigma_i}}^{u_i} \circ I_{u'}^{u_i^{\sigma_i}} \circ I_{u',i}$ (see (8.32)), which we use to define $\mathbf{e}'_{i,a,(\kappa)}(\sigma, \rho, T)$ in (8.40), is a tensor product of two $\mathbb{C}$-linear maps: one is a map between $\Lambda^{0,1}$ bundles: the other is a map between sections of the pull back bundles of TX. We denote the former by I_Σ and the latter by I_X.

I_Σ is a composition of (8.21), (8.36) and (8.38). Note this map does not depend on u' but depends only on σ and T, (which determine the source curve of u'). (8.21) and (8.38) are independent of T and depend smoothly on σ.

We apply Proposition 8.35 to $\Sigma_i^{(a)} = \Sigma_i^{\mathfrak{ob}}$, etc. and $\Sigma_i^{(b)} = \Sigma_i$ etc.. We can then apply (8.76) to estimate the map $\Psi_{i,\rho,T}$ below.

$$\Psi_{i,\rho,T} = (\mathfrak{I}_{\Xi_i}^{\sigma})^{-1} \circ \mathfrak{I}_{\Xi_i^{\mathfrak{ob}}}^{\sigma'} : K_i^{\mathfrak{ob}} \to K_i.$$

We then obtain

$$\left\| \nabla_{\sigma,\rho}^n \frac{\partial^\ell}{\partial T^\ell} \Psi_{i,\rho,T} \right\|_{C^m} \le C_{m,(8.80)} e^{-\delta_4 T}, \tag{8.80}$$

for $m \ge n$, $m \ge \ell > 0$. Note (8.72) follows from Condition 8.13 (3).

We remark that in the definition of (8.36) the process of pulling back the differential form by $\Psi_{\rho,T}$ is included. This is usually a difficult process to study in

[5] During the proof of (8.60) we apply Proposition 8.35 (2) replacing a,b by 2,1. We then change the variables by the map obtained from Proposition 8.35 (1) replacing a,b by 1,2.

Sobolev spaces. However in our situation we apply it to smooth forms $\mathbf{e}_{i,a}$ which are fixed during the construction. So we can use (8.80) to deduce the next inequality

$$\left\| \nabla^n_{\sigma,\rho} \frac{\partial^\ell}{\partial T^\ell} I_\Sigma(\mathbf{e}_{i,a}) \right\|_{C^m} \le C_{m,(8.81)} e^{-\delta_4 T}. \tag{8.81}$$

We next discuss I_X. By definition I_X is induced by the composition of the parallel transportations (8.20), (8.37) and (8.39). In the case when $u' = u^{\sigma,\rho}_{T,(\kappa)}$ we can estimate it by using induction hypothesis (that is, the version of (6.10) including σ derivatives). We obtain this estimate in the same way as in the proof of Lemma 6.21 given in Appendix F.

Thus together with (8.81) we obtain the required estimate (8.42). □

REMARK 8.36. Proposition 8.35 (1) is a variation of [**FOOO2**, Proposition 16.11] and Proposition 8.35 (2) is a variation of [**FOOO2**, Proposition 16.15]. Actually the assumptions of [**FOOO2**, Proposition 16.11] and of [**FOOO2**, Proposition 16.15] are weaker than that of Proposition 8.35. Namely in [**FOOO2**] we studied a *smooth* family of coordinates at the 0-th marked point. Here we consider an *analytic* family of coordinates at the 0-th marked point.

The proof of [**FOOO2**, Propositions 16.11 and 16.15] is given in [**FOOO2**, Section 25] and uses a method similar to the proof of Theorems 3.13 and 6.4 of this paper to find a biholomorphic map between Riemann surfaces with appropriate estimate. In other words it is based on the study of non-linear partial differential equation. The proof of Proposition 8.35 we provide in this chapter is based on the complex geometry and is shorter than [**FOOO2**, Section 25].

In case when the almost complex structure of the target space X is integrable, we may prove a similar estimate as Theorem 6.4 for the moduli space of stable maps without boundary, by using the complex geometry in a similar way as the proof of Proposition 8.35. Namely we may use the existence of the universal family of stable maps (with obstruction bundles) in the complex analytic category and translate the complex analyticity of this moduli space into an exponential decay estimate.[6]

In the situation of Proposition 8.35, the target space is not involved. So all the complex structures involved are integrable. This is the reason we can use the complex geometry to find a shorter proof. Since for the purpose of this paper (and all the applications we can see at this stage), we can always restrict ourselves to an analytic family of coordinates, we provide this shorter proof in this paper.

The existence of smooth coordinate change between Kuranishi charts in this chapter and those in [**FOOO2**] can be proved by using [**FOOO2**, Propositions 16.11 and 16.15].

In the genus 0 case, a result corresponding to Proposition 8.35 is proved in [**FOOO1**, Lemma A1.59]. The proof there uses hyperbolic geometry and is different from both of this paper and [**FOOO2**].

REMARK 8.37. As we mentioned already the proof of Proposition 8.19 is the main extra point in the proof of Theorem 8.16 other than those appearing in the proof of Theorems 3.13 and 6.4. In [**FOOO2**], this point was discussed in detail as the proof of [**FOOO2**, Lemma 19.14].

[6]Because we need to study the case when obstruction bundle is present, it is nontrivial to work out the proof of exponential decay in this way.

The contents of this chapter are taken from [**FOOO2**, Part 4]. The contents of other chapters of this paper are taken from [**FOOO2**, Part 3].

APPENDIX A

Error term estimate of non-linear Cauchy-Riemann equation I

Let Ω be an open subset of a bordered Riemann surface Σ and $u_1 : (\Omega, \Omega \cap \partial\Sigma) \to (X, L)$ a pseudoholomorphic map. Consider two smooth sections $V^0, W^0 \in \Gamma(\Omega, u_1^*TX)$ such that their restrictions to $\Omega \cap \partial\Sigma$ are in $\Gamma(\Omega \cap \partial\Sigma, u_1^*TX)$. We study the maps

$$u(z) = \mathrm{Exp}(u_1(z), W^0(z))$$

and

$$v_r(z) = \mathrm{Exp}(u(z), r\mathrm{Pal}_{u_1}^{u}(V^0)(z)). \tag{A.1}$$

We take a trivialization of u_1^*TX on Ω and identify

$$u_1^*TX \cong \Omega \times \mathbb{R}^n.$$

We write an element (of the total space of) u_1^*TX as $(z, (\xi_1, \dots, \xi_n))$. So V^0, W^0 are regarded as $(V_j^0)_{j=1}^n : \Omega \to \mathbb{R}^n$, $(W_j^0)_{j=1}^n : \Omega \to \mathbb{R}^n$. Let $z = x + \sqrt{-1}y$ be a complex coordinate of Ω.

Let $R > 0$ be a number smaller than $\iota'_X/10$. We denote by $D^n(R)$ the ball of radius R centered at 0 in $\mathbb{R}^n$.

We define $\hat{F} : \Omega \times D^n(R) \times D^n(R) \to X$ by

$$\hat{F}(z, \mathfrak{v}, \mathfrak{w}) = \mathrm{Exp}\left(\mathrm{Exp}(u_1(z), \mathfrak{w}), \mathrm{Pal}_{u_1(z)}^{\mathrm{Exp}(u_1(z),\mathfrak{w})}(\mathfrak{v})\right)$$

where $\Omega \times \mathbb{R}^n \times \mathbb{R}^n \supset \Omega \times D^n(R) \times D^n(R)$ is identified with the total space of the direct sum bundle $u_1^*(TX \oplus TX)$. See Figure A.1.

We also define $F : \Omega \times D^n(R) \times D^n(R) \to \mathbb{R}^n$ by

$$F(z, \mathfrak{v}, \mathfrak{w}) = \mathrm{E}(u_1(z), \hat{F}(z, \mathfrak{v}, \mathfrak{w})).$$

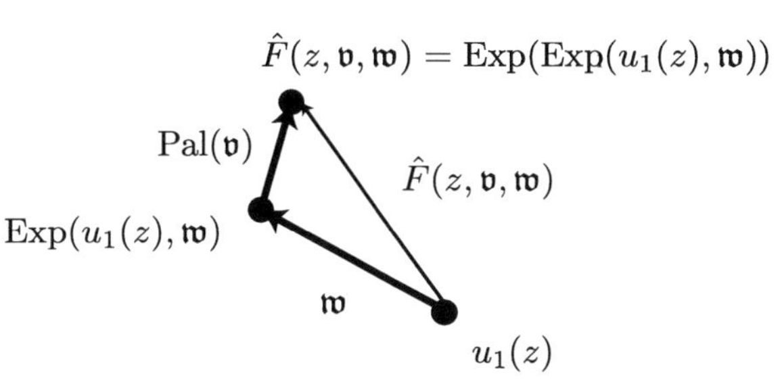

FIGURE A.1. $F(z, \mathfrak{v}, \mathfrak{w})$ and $\hat{F}(z, \mathfrak{v}, \mathfrak{w})$

We denote

$$\mathcal{PP}(z,\mathfrak{v},\mathfrak{w})=\left(\left(\mathrm{Pal}_{u_1(z)}^{\mathrm{Exp}(u_1(z),\mathfrak{w})}\right)^{(0,1)}\right)^{-1}\circ\left(\left(\mathrm{Pal}^{\hat F(z,\mathfrak{v},\mathfrak{w})}_{\mathrm{Exp}(u_1(z),\mathfrak{w})}\right)^{(0,1)}\right)^{-1}$$
$$: T_{\hat F(z,\mathfrak{v},\mathfrak{w})}X \otimes \Lambda^{0,1}\Omega \to T_{u_1(z)}X \otimes \Lambda^{0,1}\Omega.$$

We remark

$$v_r(z) = \hat F(z, rV^0(z), W^0(z)).$$

We study

$$\mathcal{PP}(z, rV^0(z), W^0(z))(\overline{\partial} v_r). \tag{A.2}$$

LEMMA A.1. *There exist smooth maps*

$$\mathfrak{G} : \Omega \times D^n(R) \times D^n(R) \to \mathbb{R}^n,$$
$$\mathfrak{H}^x_{\mathfrak{v},j}, \mathfrak{H}^x_{\mathfrak{w},j}, \mathfrak{H}^y_{\mathfrak{v},j}, \mathfrak{H}^y_{\mathfrak{w},j} : \Omega \times D^n(R) \times D^n(R) \to \mathbb{R}^n$$

$j = 1, \dots, n$ *such that*

$$\begin{aligned}(\text{A.2}) = \mathfrak{G}(z, rV^0(z), W^0(z)) &+ r\sum_j \frac{\partial V^0_j}{\partial x}(z)\mathfrak{H}^x_{\mathfrak{v},j}(z, rV^0(z), W^0) \\ &+ \sum_j \frac{\partial W^0_j}{\partial x}(z)\mathfrak{H}^x_{\mathfrak{w},j}(z, rV^0(z), W^0(z)) \\ &+ r\sum_j \frac{\partial V^0_j}{\partial y}(z)\mathfrak{H}^y_{\mathfrak{v},j}(z, rV^0(z), W^0(z)) \\ &+ \sum_j \frac{\partial W^0_j}{\partial y}(z)\mathfrak{H}^y_{\mathfrak{w},j}(z, rV^0(z), W^0(z)).\end{aligned} \tag{A.3}$$

We remark that the maps $\mathfrak{G}$, $\mathfrak{H}^x_{\mathfrak{v},j}, \mathfrak{H}^x_{\mathfrak{w},j}, \mathfrak{H}^y_{\mathfrak{v},j}, \mathfrak{H}^y_{\mathfrak{w},j}$ depend only on Ω, u_1, X and are independent of V^0, W^0.

PROOF. We emphasize that we do not need to obtain an explicit form of the smooth maps appearing in (A.3).

We first observe that there exists an $n \times n$ matrix valued smooth map H on $\Omega \times D^n(R) \times D^n(R)$ such that

$$
\begin{aligned}
\frac{\partial F(z,\mathfrak{v},\mathfrak{w})}{\partial x} &= H(z,\mathfrak{v},\mathfrak{w}) \left(\mathcal{PP}(z,\mathfrak{v},\mathfrak{w})\Big(\frac{\partial \hat F(z,\mathfrak{v},\mathfrak{w})}{\partial x}\Big)\right. \\
&\quad \left. + \frac{\partial}{\partial x}\mathrm{E}(u_1(z),\xi)\Big|_{\xi=\hat F(z,\mathfrak{v},\mathfrak{w})}\right) \\
\frac{\partial F(z,\mathfrak{v},\mathfrak{w})}{\partial y} &= H(z,\mathfrak{v},\mathfrak{w}) \left(\mathcal{PP}(z,\mathfrak{v},\mathfrak{w})\Big(\frac{\partial \hat F(z,\mathfrak{v},\mathfrak{w})}{\partial y}\Big)\right. \\
&\quad \left. + \frac{\partial}{\partial y}\mathrm{E}(u_1(z),\xi)\Big|_{\xi=\hat F(z,\mathfrak{v},\mathfrak{w})}\right) \\
\frac{\partial F(z,\mathfrak{v},\mathfrak{w})}{\partial \mathfrak{v}_j} &= H(z,\mathfrak{v},\mathfrak{w}) \left(\mathcal{PP}(z,\mathfrak{v},\mathfrak{w})\Big(\frac{\partial \hat F(z,\mathfrak{v},\mathfrak{w})}{\partial \mathfrak{v}_j}\Big)\right) \\
\frac{\partial F(z,\mathfrak{v},\mathfrak{w})}{\partial \mathfrak{w}_j} &= H(z,\mathfrak{v},\mathfrak{w}) \left(\mathcal{PP}(z,\mathfrak{v},\mathfrak{w})\Big(\frac{\partial \hat F(z,\mathfrak{v},\mathfrak{w})}{\partial \mathfrak{w}_j}\Big)\right).
\end{aligned} \tag{A.4}
$$

In fact we have

$$H(z,\mathfrak{v},\mathfrak{w}) = (D_{\hat F(z,\mathfrak{v},\mathfrak{w}).}\mathrm{E}(u_1(z),\cdot))^{-1} \circ \mathcal{PP}(z,\mathfrak{v},\mathfrak{w})^{-1}.$$

Here $(D_{\hat F(z,\mathfrak{v},\mathfrak{w})}\mathrm{E})(u_1(z),\cdot)$ is the differential of the map $\xi \mapsto \mathrm{E}(u_1(z),\xi)$ at $\xi = \hat F(z,\mathfrak{v},\mathfrak{w}) \in X$.

We next observe that there exists a matrix valued smooth function $\mathfrak{J}$ on $\Omega \times D^n(R) \times D^n(R)$ such that

$$\mathfrak{J}(z,\mathfrak{v},\mathfrak{w})(\mathcal{PP}(z,\mathfrak{v},\mathfrak{w})(\xi)) = \mathcal{PP}(z,\mathfrak{v},\mathfrak{w})(J_{\hat F(z,\mathfrak{v},\mathfrak{w})}(\xi)) \tag{A.5}$$

for any $\xi \in T_{\hat F(z,\mathfrak{v},\mathfrak{w})}X$. Here $J_{\hat F(z,\mathfrak{v},\mathfrak{w})} : T_{\hat F(z,\mathfrak{v},\mathfrak{w})}X \to T_{\hat F(z,\mathfrak{v},\mathfrak{w})}X$ is the almost complex structure of X.

By definition

$$\overline{\partial} v_r = \frac{\partial}{\partial x}\hat F(z,rV^0(z),W^0(z)) + J_{\hat F(z,rV^0(z),W^0(z))}\left(\frac{\partial}{\partial y}\hat F(z,rV^0(z),W^0(z))\right).$$

Note

$$
\begin{aligned}
\frac{\partial}{\partial x}\hat F(z,rV^0(z),W^0(z)) &= \frac{\partial \hat F}{\partial x} + \sum_j r\frac{\partial V^0_j}{\partial x}\frac{\partial \hat F}{\partial \mathfrak{v}_j} + \sum_j \frac{\partial W^0_j}{\partial x}\frac{\partial \hat F}{\partial \mathfrak{w}_j} \\
\frac{\partial}{\partial y}\hat F(z,rV^0(z),W^0(z)) &= \frac{\partial \hat F}{\partial y} + \sum_j r\frac{\partial V^0_j}{\partial y}\frac{\partial \hat F}{\partial \mathfrak{v}_j} + \sum_j \frac{\partial W^0_j}{\partial y}\frac{\partial \hat F}{\partial \mathfrak{w}_j}.
\end{aligned}
$$

Now it is fairly obvious that we can find $\mathfrak{G}, \mathfrak{H}^x_{\mathfrak{v},j}, \mathfrak{H}^x_{\mathfrak{w},j}, \mathfrak{H}^y_{\mathfrak{v},j}, \mathfrak{H}^y_{\mathfrak{w},j}$ from H, $\mathfrak{J}$, the x,y derivatives of $\mathrm{E}(u_1(z),\xi)$ with $\xi = \hat F$, and

$$
\begin{aligned}
&\mathcal{PP}(z,\mathfrak{v},\mathfrak{w})\left(\frac{\partial \hat F}{\partial x}\right), && \mathcal{PP}(z,\mathfrak{v},\mathfrak{w})\left(\frac{\partial \hat F}{\partial y}\right), \\
&\mathcal{PP}(z,\mathfrak{v},\mathfrak{w})\left(\frac{\partial \hat F}{\partial \mathfrak{v}_j}(z,\mathfrak{v},\mathfrak{w})\right), && \mathcal{PP}(z,\mathfrak{v},\mathfrak{w})\left(\frac{\partial \hat F}{\partial \mathfrak{w}_j}(z,\mathfrak{v},\mathfrak{w})\right).
\end{aligned} \tag{A.6}
$$

(A.6) are also $\mathbb{R}^n$ valued smooth functions of $(z, \mathfrak{v}, \mathfrak{w})$ because of (A.4).

The proof of Lemma A.1 is complete. □

By (A.3) we have

$$\begin{aligned}\frac{d^2}{dr^2}(\text{A.2}) &= \sum_{ij} V_i^0 V_j^0 \mathfrak{F}_{ij}\left(r, z, W^0, V^0, \frac{\partial W^0}{\partial x}, \frac{\partial W^0}{\partial y}, \frac{\partial V^0}{\partial x}, \frac{\partial V^0}{\partial y}\right) \\ &+ \sum_{ij} V_i^0 \frac{\partial V_j^0}{\partial x} \mathfrak{F}^x_{ij}(r, z, W^0, V^0) + \sum_{ij} V_i^0 \frac{\partial V_j^0}{\partial y} \mathfrak{F}^y_{ij}(r, z, W^0, V^0)\end{aligned} \tag{A.7}$$

where $\mathfrak{F}_{ij}$, $\mathfrak{F}^x_{ij}$, $\mathfrak{F}^y_{ij}$ are smooth maps independent of V^0,W^0.

Proof of Lemma 5.19. We take: $\Sigma = \Sigma_1$, and Ω_α $(\alpha = 1, \dots, \mathfrak{A})$ open subsets of Σ_1 such that $\bigcup_{\alpha=1}^{\mathfrak{A}} \Omega_\alpha \supset K_1$. We also put $u_1 = u_1$ and

$$\begin{aligned}W^0(z) &= \text{E}(u_1(z), u(z)) \in T_{u_1(z)}(X), \\ V^0(z) &= \text{Pal}_{u(z)}^{u_1(z)}(V(z)) \in T_{u_1(z)}(X).\end{aligned}$$

Then by (A.7), we obtain:

$$\left\| \int_0^1 ds \int_0^s \frac{d^2}{dr^2}(\text{A.2})\, dr \right\|_{L^2_m(\Omega_\alpha)} \le C_{(\text{A.8})} \left\| V^0 \right\|^2_{L^2_{m+1}(\Omega_\alpha)} \tag{A.8}$$

where $C_{(\text{A.8})}$ depends on $\max\{\|V^0\|_{L^2_{m+1}(\Omega_\alpha)}, \|W^0\|_{L^2_{m+1}(\Omega_\alpha)}\}$.

By taking the sum over α, we obtain the inequality (5.44) to be proven. □

Proof of (5.59). By putting $\Sigma = \Sigma(S)$ the proof is the same as the proof of Lemma 5.19. □

Proof of the first inequality of (5.64). We put $\Sigma = \Sigma_1$, Ω a neighborhood of $\mathcal{A}_T$, $u_1 = u_1$,

$$\begin{aligned}u(z) &= \text{Exp}(u^\rho_{T,0}(z), V^\rho_{T,1,(1)}), \\ v_r(z) &= \text{Exp}(u^\rho_{T,0}(z), V^\rho_{T,1,(1)} + r\chi_{\vec{\mathcal{A}}}(V^\rho_{T,2,(1)} - (\Delta p^\rho_{T,(1)})^{\text{Pal}})).\end{aligned}$$

We then define

$$\begin{aligned}W^0(z) &= \text{E}(u_1(z), u(z)) \in T_{u_1(z)}(X), \\ V^0(z) &= \text{Pal}_{u(z)}^{u_1(z)}(\chi_{\vec{\mathcal{A}}}(V^\rho_{T,2,(1)} - (\Delta p^\rho_{T,(1)})^{\text{Pal}})) \in T_{u_1(z)}(X).\end{aligned}$$

We have

$$\mathcal{P}\overline{\partial} v_{r_0} - \overline{\partial} u = \int_0^{r_0} \left(\frac{\partial}{\partial r}\right) \mathcal{P}(\overline{\partial} v_r) dr \tag{A.9}$$

Here $\mathcal{P}$ is the inverse of the complex linear part of the parallel transport along the path $r \mapsto v_r(z)$. This path is not a geodesic. However we can apply the same argument as the proof of Lemma A.1 to obtain

$$
\begin{aligned}
\mathfrak{P}\left(\frac{\partial}{\partial r}\right)\mathcal{P}(\overline{\partial} v_r) = & \sum_i V_i^0 \mathfrak{F}_i\left(r, z, W^0, V^0, \frac{\partial W^0}{\partial x}, \frac{\partial W^0}{\partial y}, \frac{\partial V^0}{\partial x}, \frac{\partial V^0}{\partial y}\right) \\
& + \sum_i \frac{\partial V_i^0}{\partial x} \mathfrak{F}_i^x\left(r, z, W^0, V^0\right) \\
& + \sum_i \frac{\partial V_i^0}{\partial y} \mathfrak{F}_i^y\left(r, z, W^0, V^0\right),
\end{aligned} \tag{A.10}
$$

where $\mathfrak{F}_i$, $\mathfrak{F}_i^x$, $\mathfrak{F}_i^y$ are smooth maps. Here $\mathfrak{P} = ((\mathrm{Pal}_{u_1}^{u})^{(0,1)})^{-1}$. The first inequality of (5.64) follows from (A.9) and (A.10). □

REMARK A.2. In the argument of this appendix or anywhere in this paper we *never* use the fact that we take the parallel transport with respect to the Levi-Civita connection. We can actually use any linear connection which preserves $TL \subset TX$ in place of the Levi-Civita connection of the metric given in Lemma 2.1, as we did in [**FOOO1**, Chapter 7].

In the case when there are several (finitely many) Lagrangian submanifolds L_α so that they have mutually clean intersection, we cannot generalize Lemma 2.1 to find a Hermitian metric such that all the L_α are totally geodesic. However it is easy to see that there exists a linear connection which preserves all the subspaces $TL_\alpha \subset TX$.

APPENDIX B

Estimate of Parallel transport 1

Let X be a Riemannian manifold. We put

$$\mathfrak{U} = \{(x,y,\mathfrak{v},\mathfrak{w}) \mid x,y \in X, d(x,y) \le \iota'_X/2, \mathfrak{v},\mathfrak{w} \in T_xX, |\mathfrak{v}|,|\mathfrak{w}| < \iota'_X/10\}, \tag{B.1}$$

which is a smooth manifold. We consider a smooth vector bundle $\mathfrak{L}$ on $\mathfrak{U}$ whose fiber at $(x,y,\mathfrak{v},\mathfrak{w})$ is

$$\mathfrak{L}_{(x,y,\mathfrak{v},\mathfrak{w})} = \mathrm{Hom}_{\mathbb{R}}(T_yX, T_xX). \tag{B.2}$$

We define its smooth section $\mathfrak{P}$ whose value at $(x,y,\mathfrak{v},\mathfrak{w})$ is as follows. We put

$$\mathfrak{p} = \mathrm{Exp}(x,\mathfrak{w}), \qquad \mathfrak{q} = \mathrm{Exp}(\mathfrak{p}, \mathrm{Pal}_x^{\mathfrak{p}}(\mathfrak{v})),$$

then

$$\mathfrak{P}(x,y,\mathfrak{v},\mathfrak{w}) = ((\mathrm{Pal}_x^{\mathfrak{p}})^J)^{-1} \circ ((\mathrm{Pal}_{\mathfrak{p}}^{\mathfrak{q}})^J)^{-1} \circ (\mathrm{Pal}_y^{\mathfrak{q}})^J. \tag{B.3}$$

The smoothness of $\mathfrak{P}$ is obvious from the definition.

LEMMA B.1. *Let $\mathscr{S}$ be a 2 dimensional manifold and $\mathfrak{v}(s)$ and $\mathfrak{w}(s)$ be an $s \in \mathcal{S}$ parameterized family of smooth maps $\mathscr{S} \to T_xX$ with $|\mathfrak{v}(s)|, |\mathfrak{w}(s)| \le i'_X/10$. If*

$$\left\| \frac{\partial^{n_1}}{\partial s^{n_1}} \mathfrak{v} \right\|_{L^2_m(\mathscr{S})} \le 1, \qquad \left\| \frac{\partial^{n_2}}{\partial s^{n_2}} \mathfrak{w} \right\|_{L^2_m(\mathscr{S})} \le 1$$

for $n_1, n_2 \le n$, then

$$\begin{aligned} &\left\| \frac{\partial^n}{\partial s^n} \Big(\mathfrak{P}(x,y,\mathfrak{v},\mathfrak{w}) - \mathfrak{P}(x,y,\mathfrak{v},0) \Big) \right\|_{L^2_m(\mathscr{S})} \\ &\le C_{m,(\mathrm{B.4})} \sum_{n' \le n} \left\| \frac{\partial^{n'}}{\partial s^{n'}} \mathfrak{w} \right\|_{L^2_m(\mathscr{S})}. \end{aligned} \tag{B.4}$$

PROOF. The lemma is an immediate consequence of the smoothness of $\mathfrak{P}$. □

PROOF OF LEMMA 5.20. We consider $\mathfrak{e} = \mathfrak{e}^{\rho}_{1,T,(0)} \in \mathcal{E}_1$. When we defined $\mathcal{E}_1$ it was a subset $\mathcal{E}_1 \subset L^2_m(\Sigma_1; (u_1^{\mathfrak{ob}})^*TX \otimes \Lambda^{0,1})$. We used the parallel transport to regard

$$\mathfrak{e} \in L^2_m(\Sigma_1; (\hat u^{\rho}_{1,T,(0)})^*TX \otimes \Lambda^{0,1}) = L^2_m(\Sigma_1; u^*TX \otimes \Lambda^{0,1}).$$

It appears in (5.47). So more precisely $\mathfrak{e}$ in (5.47) is:

$$z \mapsto (\mathrm{Pal}_{u_1^{\mathfrak{ob}}(z)}^{\hat u^{\rho}_{1,T,(0)}(z)})^{(0,1)}(\mathfrak{e}^{\rho}_{1,T,(0)}(z)).$$

(Here $z \in K_1^{\mathfrak{ob}}$.) The expression $\mathcal{P}^{-1}\mathfrak{e}$ in (5.47) is then equal to

$$z \mapsto (\mathrm{Pal}_{\hat u^{\rho}_{1,T,(0)}(z)}^{\mathrm{Exp}(\hat u^{\rho}_{1,T,(0)}(z), V^{\rho}_{T,1,(1)}(z))})^{(0,1)} \circ (\mathrm{Pal}_{u_1^{\mathfrak{ob}}(z)}^{\hat u^{\rho}_{1,T,(0)}(z)})^{(0,1)}(\mathfrak{e}^{\rho}_{1,T,(0)}(z)). \tag{B.5}$$

On the other hand $\mathfrak{e} \in L^2_m(\Sigma_i; \mathrm{Exp}(\hat{u}^{\rho}_{1,T,(0)}, V^{\rho}_{T,1,(1)})^*TX \otimes \Lambda^{0,1})$ appearing in Lemma 5.20 is, by definition,

$$z \mapsto (\mathrm{Pal}_{u_1^{\mathfrak{ob}}(z)}^{\mathrm{Exp}(\hat{u}^{\rho}_{T,1,(0)}(z), V^{\rho}_{T,1,(1)}(z))})^{(0,1)}(\mathfrak{e}^{\rho}_{1,T,(0)}(z)). \tag{B.6}$$

Using the section $\mathfrak{P}$ in (B.3), the sections (B.5) and (B.6) are written as follows.

$$\begin{aligned}(\mathrm{B.5}) &= \mathfrak{P}(\hat{u}^{\rho}_{T,1,(0)}(z), u_i^{\mathfrak{ob}}(z), 0, V^{\rho}_{T,1,(1)}(z)))(\mathfrak{e}^{\rho}_{1,T,(0)}(z)),\\ (\mathrm{B.6}) &= \mathfrak{P}(\hat{u}^{\rho}_{T,1,(0)}(z), u_i^{\mathfrak{ob}}(z), 0, 0))(\mathfrak{e}^{\rho}_{1,T,(0)}(z)).\end{aligned}$$

Therefore (B.5)-(B.6) are estimated by $Ce^{-\delta_1 T}$ by Lemma B.1. Then the estimate (5.57) easily follows from this. $\square$

APPENDIX C

Error term estimate of non-linear Cauchy-Riemann equation II

In this appendix, we give a proof of Proposition 5.34. Certain estimates of the parallel transport are postponed till the next appendix.

PROOF OF PROPOSITION 5.34. The proof is similar to that of Proposition 5.17 and proceeds as follows.

We first perform the estimates on K_1. We use the simplified notation:

$$\begin{aligned} &u = \hat{u}^{\rho}_{1,T,(\kappa-1)}, && V = V^{\rho}_{T,1,(\kappa)}, \\ &\mathcal{P} = (\mathrm{Pal}_1^{(0,1)})^{-1}, && \mathfrak{e} = \sum_{a=0}^{\kappa-1} \mathfrak{e}^{\rho}_{1,T,(a)}. \end{aligned} \tag{C.1}$$

Here $\mathrm{Pal}_1^{(0,1)}$ is the $(0,1)$ part of the parallel translation along the map $s \mapsto \mathrm{Exp}(u, sV)$, $s \in [0, s_0]$ for some $s_0 \in [0,1]$.

We obtain the same formula as (5.43)

$$\begin{aligned} &\mathcal{P}\overline{\partial}(\mathrm{Exp}(u,V)) \\ &= \overline{\partial}(\mathrm{Exp}(u,0)) + \int_0^1 \frac{\partial}{\partial s}\left(\mathcal{P}\overline{\partial}(\mathrm{Exp}(u,sV))\right) ds \\ &= \overline{\partial}(\mathrm{Exp}(u,0)) + (D_u\overline{\partial})(V) + \int_0^1 ds \int_0^s \left(\frac{\partial}{\partial r}\right)^2 (\mathcal{P}\overline{\partial}(\mathrm{Exp}(u,rV))\, dr. \end{aligned} \tag{C.2}$$

Then we also obtain

$$\begin{aligned} &\left\Vert \int_0^1 ds \int_0^s \left(\frac{\partial}{\partial r}\right)^2 (\mathcal{P}\overline{\partial}(\mathrm{Exp}(u,rV))\, dr \right\Vert_{L^2_m(K_1)} \\ &\qquad \le C_{m,(\mathrm{C.3})} \Vert V \Vert^2_{L^2_{m+1,\delta}} \le C'_{m,(\mathrm{C.3})} e^{-2\delta_1 T} \mu^{2(\kappa-1)} \end{aligned} \tag{C.3}$$

provided $m \ge 2$. One can prove (C.3) by applying Lemma A.1 to

$$W^0(z) = \mathrm{E}(u_1(z), \hat{u}^{\rho}_{1,T,(\kappa-1)}(z)), \quad V^0(z) = \mathrm{Pal}^{u_1(z)}_{\hat{u}^{\rho}_{1,T,(\kappa-1)}(z)} \left(V^{\rho}_{T,1,(\kappa)}(z)\right) \tag{C.4}$$

and applying the Sobolev inequality etc. to the right hand side of (A.7).

REMARK C.1. We remark that we are working by induction on κ and take infinitely many steps $\kappa = 1, 2, \dots$. For this proof to work we need the constants $C_{m,(\mathrm{C.3})}$ and $C'_{m,(\mathrm{C.3})}$ in (C.3) to be independent of κ.

The reason we can take $C_{m,(\mathrm{C.3})}$ and $C'_{m,(\mathrm{C.3})}$ to be independent of κ is as follows. As we mentioned right after (A.8) the constant $C_{(\mathrm{A.8})}$ there depends only on $\max\{\Vert V^0 \Vert_{L^2_{m+1}(\Omega)}, \Vert W^0 \Vert_{L^2_{m+1}(\Omega)}\}$. In our situation where W^0 and V^0 are given by (C.4), their local L^2_{m+1} norms are bounded by the induction hypothesis ((6.9) and (6.10)) and Lemma 6.12, by a number independent of κ.

Therefore $C_{m,(\mathrm{C}.3)}$ and $C'_{m,(\mathrm{C}.3)}$ can be taken to be independent of κ. It can be taken to be independent of T since we are working on K_1 which is independent of T.

Independence of the constants of κ or T appears in other part of the proof, which can be proved in the same way. So we do not mention it usually.

Next we have

$$\begin{aligned}
&\mathcal{P} \circ \Pi^{\perp}_{\mathcal{E}_1(\mathrm{Exp}(u,V))} \circ \mathcal{P}^{-1} \\
&= \Pi^{\perp}_{\mathcal{E}_1(u)} + \int_0^1 \frac{d}{ds}\left(\mathcal{P} \circ \Pi^{\perp}_{\mathcal{E}_1(\mathrm{Exp}(u,sV)} \circ \mathcal{P}^{-1}\right) ds \\
&= \Pi^{\perp}_{\mathcal{E}_1(u)} - (D_u\mathcal{E}_1)(\cdot, V) + \int_0^1 ds \int_0^s \frac{d^2}{dr^2}\left(\mathcal{P} \circ \Pi^{\perp}_{\mathcal{E}_1(\mathrm{Exp}(u,rV))} \circ \mathcal{P}^{-1}\right) dr,
\end{aligned} \tag{C.5}$$

in which we can estimate the third term in the same way as (C.3). (See Appendix D.)

On the other hand, (5.91) for $\le \kappa - 1$, (5.68) for $\le \kappa$ and (C.2), (C.3) imply

$$\left\|\overline{\partial}(\mathrm{Exp}(u,V)) - \mathcal{P}^{-1}\mathfrak{e}\right\|_{L^2_m(K_1)} \le C_{m,(\mathrm{C}.6)} e^{-\delta_1 T}\mu^{\kappa-1}. \tag{C.6}$$

We also use the next inequality

$$\left\|\mathcal{P}^{-1}(D_u\mathcal{E}_1)(\mathcal{P}Q, V)\right\|_{L^2_m(K_1)} \le C_{m,(\mathrm{C}.7)} \|Q\|_{L^2_m(K_1)} \|V\|_{L^2_m(K_1)}. \tag{C.7}$$

Here Q is a section of $u^*TX \otimes \Lambda^{0,1}$ of L^2_m class. (C.7) is a version of (5.49) and is proved in Appendix D. Using (C.6) and (C.7) we can show

$$\begin{aligned}
&\Big\|\Pi^{\perp}_{\mathcal{E}_1(\mathrm{Exp}(u,V))}\overline{\partial}(\mathrm{Exp}(u,V)) - \mathcal{P}^{-1}\Pi^{\perp}_{\mathcal{E}_1(u)}(\mathcal{P}\overline{\partial}(\mathrm{Exp}(u,V)) \\
&\qquad + \mathcal{P}^{-1}(D_u\mathcal{E}_1)(\mathfrak{e}, V)\Big\|_{L^2_m(K_1^+)} \le C_{m,(\mathrm{C}.8)} e^{-2\delta_1 T}\mu^{\kappa-1},
\end{aligned} \tag{C.8}$$

in the same way as (5.52).

By (5.85) we have:

$$\overline{\partial}u + (D_u\overline{\partial})(V) - (D_u\mathcal{E}_1)(\mathfrak{e}, V) \in \mathcal{E}_1(u) \tag{C.9}$$

on K_1.

Summing up we have derived

$$\begin{aligned}
\|\Pi^{\perp}_{\mathcal{E}_1(\mathrm{Exp}(u,V))}(\overline{\partial}(\mathrm{Exp}(u,V)))\|_{L^2_m(K_1)} &\le C_{m,(\mathrm{C}.10)} e^{-2\delta_1 T}\mu^{\kappa-1} \\
&\le C_{1,m} e^{-\delta_1 T}\epsilon(5)\mu^{\kappa}/10
\end{aligned} \tag{C.10}$$

for $T > T_{m,\epsilon(5),(\mathrm{C}.10)}$, as long as we choose $T_{m,\epsilon(5),(\mathrm{C}.10)}$ so that

$$C_{m,(\mathrm{C}.10)} e^{-\delta_1 T_{m,\epsilon(5),(\mathrm{C}.10)}} \le C_{1,m}\epsilon(5)\mu/10.$$

We emphasize that this choice of $T_{m,\epsilon(5),(\mathrm{C}.10)}$ does not depend on κ but depends only on m, $\epsilon(5)$ and μ.

It follows from (C.6) that

$$\|\Pi_{\mathcal{E}_1(\mathrm{Exp}(u,V))}\left(\overline{\partial}(\mathrm{Exp}(u,V))\right) - \mathfrak{e}\|_{L^2_m(K_1)} \le C_{m,(\mathrm{C}.11)} e^{-\delta T}\mu^{\kappa-1}, \tag{C.11}$$

in the same way as the proof of Lemma 5.20 given in Appendix B.

Then we can prove (5.92) by putting

$$\mathfrak{e}^{\rho}_{1,T,(\kappa)} = \Pi_{\mathcal{E}_1(\mathrm{Exp}(u,V))}(\overline{\partial}(\mathrm{Exp}(u,V)) - \mathfrak{e} \in \mathcal{E}_1(\mathrm{Exp}(u,V)) \cong \mathcal{E}_1^{\mathfrak{ob}}. \tag{C.12}$$

We can perform the estimate on $[-5T, -T-1] \times [0,1]$ modifying the proof of (5.61) in the same way as follows. For $[S, S+1] \times [0,1] \subset [-5T, -T-1]_\tau \times [0,1]$ the inequality

$$\left\| \overline{\partial} u^{\rho}_{T,(\kappa)} \right\|_{L^2_m([S,S+1]\times[0,1]\subset\Sigma_T)} < C_{m,(\mathrm{C.13})}\mu^{\kappa-1}e^{-2\delta_1 T} \tag{C.13}$$

can be proved in the same way as (C.10). Therefore

$$\begin{aligned} \left\| \overline{\partial} u^{\rho}_{T,(\kappa)} \right\|_{L^2_{m,\delta}([-5T,-T-1]\times[0,1]\subset\Sigma_T)} &\le 40Te^{4T\delta}C_{m,(\mathrm{C.13})}\mu^{\kappa-1}e^{-2\delta_1 T} \\ &\le C_{1,m}e^{-\delta_1 T}\epsilon(5)\mu^{\kappa}/10, \end{aligned} \tag{C.14}$$

for $T > T_{m,\epsilon(5),(\mathrm{C.14})}$. (Here we use the fact that the weight function $e_{T,\delta}$ is smaller than $10e^{4T\delta}$ on our domain.)

The estimates on K_2 and $[T+1, 5T] \times [0,1]$ are the same. Notation (C.1) is used up to here.

The estimate $\overline{\partial} u^{\rho}_{T,(\kappa)}$ on $[-T+1, T-1] \times [0,1]$ is as follows. Note the bump functions $\chi^{\leftarrow}_{\mathcal{B}}$ and $\chi^{\rightarrow}_{\mathcal{A}}$ are $\equiv 1$ there and

$$\overline{\partial} u^{\rho}_{T,(\kappa-1)} + (D_{u^{\rho}_{T,(\kappa-1)}}\overline{\partial})(V^{\rho}_{T,1,(\kappa)} + V^{\rho}_{T,2,(\kappa)}) = 0.$$

Therefore the inequality

$$\left\| \overline{\partial} u^{\rho}_{T,(\kappa)} \right\|_{L^2_m([-T+1,T-1]\times[0,1]\subset\Sigma_T)} < C_{m,(\mathrm{C.15})}e^{-2\delta_1 T}\mu^{\kappa-1} \tag{C.15}$$

can be proved in the same way as (C.10). Since $e_{T,\delta} \le 10e^{5\delta T} \le 10e^{\delta_1 T/2}$, it implies

$$\left\| \overline{\partial} u^{\rho}_{T,(\kappa)} \right\|_{L^2_{m,\delta}([-T+1,T-1]\times[0,1]\subset\Sigma_T)} < e^{-\delta_1 T}\epsilon(5)\mu^{\kappa}/10 \tag{C.16}$$

for $T > T_{m,\epsilon(5),(\mathrm{C.16})}$.

On $\mathcal{A}_T$ we have

$$u^{\rho}_{T,(\kappa)} = \mathrm{Exp}(u^{\rho}_{T,(\kappa-1)}, \chi^{\rightarrow}_{\mathcal{A}}(V^{\rho}_{T,2,(\kappa)} - \Delta p^{\rho}_{T,(\kappa)}) + V^{\rho}_{T,1,(\kappa)}). \tag{C.17}$$

Note

$$\begin{aligned} &\|\chi^{\rightarrow}_{\mathcal{A}}(V^{\rho}_{T,2,(\kappa)} - \Delta p^{\rho}_{T,(\kappa)})\|_{L^2_{m+1}(\mathcal{A}_T)} \\ &\le C_{m,(\mathrm{C.18})}e^{-6T\delta}\|V^{\rho}_{T,2,(\kappa)} - \Delta p^{\rho}_{T,(\kappa)}\|_{L^2_{m+1,\delta}(\mathcal{A}_T\subset\Sigma_2)} \\ &\le C'_{m,(\mathrm{C.18})}e^{-6T\delta-T\delta_1}\mu^{\kappa-1}. \end{aligned} \tag{C.18}$$

The first inequality follows from the fact that the weight function $e_{2,\delta}$ is around $e^{6T\delta}$ on $\mathcal{A}_T$. The second inequality follows from (5.87). On the other hand the weight function $e_{T,\delta}$ is around $e^{4T\delta}$ at $\mathcal{A}_T$. (C.18) implies

$$\|\chi^{\rightarrow}_{\mathcal{A}}(V^{\rho}_{T,2,(\kappa)} - \Delta p^{\rho}_{T,(\kappa)})\|_{L^2_{m+1,\delta}(\mathcal{A}_T\subset\Sigma_T)} \le C_{m,(\mathrm{C.19})}e^{-2T\delta-T\delta_1}\mu^{\kappa-1}. \tag{C.19}$$

Therefore in the same way as the proof of (5.64) given at the end of Appendix A, we obtain

$$\begin{aligned}&\left|\|\overline{\partial}u^{\rho}_{T,(\kappa-1)}\|_{L^2_{m,\delta}(\mathcal{A}_T\subset\Sigma_T)} - \|\overline{\partial}(\mathrm{Exp}(u^{\rho}_{T,(\kappa-1)},V^{\rho}_{T,1,(\kappa)})\|_{L^2_{m,\delta}(\mathcal{A}_T\subset\Sigma_T)}\right| \\ &\le C_{m,(\mathrm{C.20})}e^{-2\delta T-\delta_1 T} \le C_{1,m}\epsilon(5)e^{-\delta_1 T}\mu^{\kappa}/20\end{aligned} \tag{C.20}$$

for $T > T_{m,\epsilon(5),(\mathrm{C.20})}$.

Since $\mathrm{Err}^{\rho}_{2,T,(\kappa-1)} = 0$ on $\mathcal{A}_T$ we have

$$\overline{\partial}u^{\rho}_{T,(\kappa-1)} + (D_{u^{\rho}_{T,(\kappa-1)}}\overline{\partial})(V^{\rho}_{T,1,(\kappa)}) = 0. \tag{C.21}$$

Therefore in the same way as we did on K_1 we can show

$$\begin{aligned}\|\overline{\partial}(\mathrm{Exp}(u^{\rho}_{T,(\kappa-1)},V^{\rho}_{T,1,(\kappa)})\|_{L^2_{m,\delta}(\mathcal{A}_T\subset\Sigma_T)} &\le C_{m,(\mathrm{C.22})}\mu^{2(\kappa-1)}e^{4\delta T}e^{-2\delta_1 T} \\ &\le C_{1,m}\epsilon(5)e^{-\delta_1 T}\mu^{\kappa}/20\end{aligned} \tag{C.22}$$

for $T > T_{m,\epsilon(5),(\mathrm{C.22})}$. (Here we use the fact that $e_{T,\delta} \sim e^{4T\delta}$ on $\mathcal{A}_T$.)

(C.20) and (C.22) imply

$$\|\overline{\partial}u^{\rho}_{T,(\kappa)}\|_{L^2_m(\mathcal{A}_T\subset\Sigma_T)} < C_{1,m}\epsilon(5)e^{-\delta_1 T}\mu^{\kappa}/10 \tag{C.23}$$

The estimate on $\mathcal{B}_T$ is similar. The proof of Proposition 5.34 is complete except the estimate of the parallel transport given in the next appendix. □

APPENDIX D

Estimate of Parallel transport 2

PROOF OF (C.7). Let Ω be a small neighborhood of K_1. We consider the vector bundle $\mathfrak{L}$ on $\mathfrak{U}$ defined by (B.1) and (B.2). We put

$$\mathfrak{V} = \{(z, \mathfrak{v}, \mathfrak{w}) \mid z \in \Omega,\ \mathfrak{v}, \mathfrak{w} \in T_{u_1(z)}X,\ |\mathfrak{v}|, |\mathfrak{w}| < R\},$$

with $R < \iota'_X/10$. We pull $\mathfrak{L}$ back by the map $\mathfrak{V} \to \mathfrak{U}$ defined by

$$(z, \mathfrak{v}, \mathfrak{w}) \mapsto (u_1(z), u_1^{\mathfrak{ob}}(z), \mathfrak{v}, \mathfrak{w})$$

to obtain a vector bundle $\hat{\mathfrak{L}}$ on $\mathfrak{V}$.

We pull back the section $\mathfrak{P}$ in (B.3) by this map and tensor it with the identity in $\mathrm{End}(\Lambda^{0,1}(\Omega))$. We then obtain a section P of $\hat{\mathfrak{L}} \otimes \mathrm{End}(\Lambda^{0,1}(\Omega))$. P is written as:

$$P(z, \mathfrak{v}, \mathfrak{w}) = \left((\mathrm{Pal}_{u_1(z)}^{\mathrm{Exp}(u_1(z),\mathfrak{w})})^{(0,1)}\right)^{-1} \circ \left((\mathrm{Pal}_{\mathrm{Exp}(u_1(z),\mathfrak{w})}^{\hat{v}(z,\mathfrak{v},\mathfrak{w})})^{(0,1)}\right)^{-1} \circ (\mathrm{Pal}_{u_1^{\mathfrak{ob}}(z)}^{\hat{v}(z,\mathfrak{v},\mathfrak{w})})^{(0,1)}$$

where

$$\hat{v}(z, \mathfrak{v}, \mathfrak{w}) = \mathrm{Exp}(\mathrm{Exp}(u_1(z), \mathfrak{w}), \mathrm{Pal}_{u_1(z)}^{\mathrm{Exp}(u_1(z),\mathfrak{w})}(\mathfrak{v})). \tag{D.1}$$

See Figure D.1.

Let $\hat{\mathbf{e}}_i(z)$ ($i = 1, \dots, \dim \mathcal{E}_1$) be a basis of $\mathcal{E}_1 = \mathcal{E}_1(u_1^{\mathfrak{ob}})$. We define

$$\hat{\mathbf{e}}_i(z, \mathfrak{v}, \mathfrak{w}) = P(z, \mathfrak{v}, \mathfrak{w})(\hat{\mathbf{e}}_i(z)),$$

which is a smooth section of the bundle $u_1^*TX \otimes \Lambda^{0,1}(\Omega)$ on $\mathfrak{V}$. (Here we denote the map of $(z, \mathfrak{v}, \mathfrak{w}) \mapsto u_1(z)$ by u_1 by a slight abuse of notation.) We consider

$$\begin{aligned} \mathbf{e}'_i(z, r) &= P(z, rV^0(z), W^0(z))(\hat{\mathbf{e}}_i(z)) \\ &= \hat{\mathbf{e}}_i(z, rV^0(z), W^0(z)) \in T_{u_1(z)}X \otimes \Lambda^{0,1}(\Omega). \end{aligned} \tag{D.2}$$

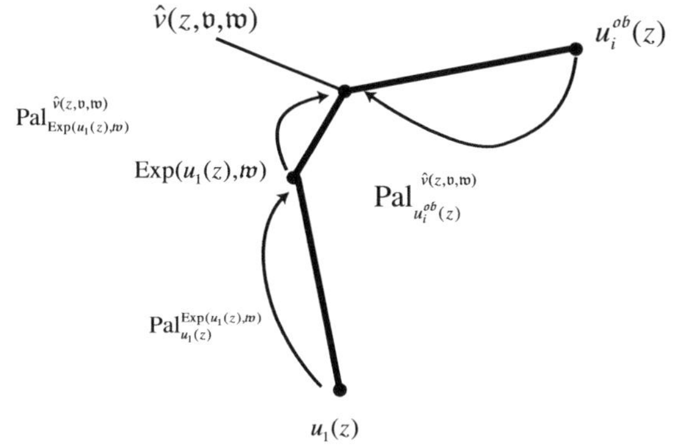

FIGURE D.1. Map P

The sections $\mathbf{e}'_i \in \Gamma(u_1^*TX \otimes \Lambda^{0,1}(\Omega))$, $(i = 1, \dots, \dim \mathcal{E}_1)$ form a basis of $P(\mathcal{E}_1)$.

In the next step, we will use the Gram-Schmidt process to modify it to an orthonormal basis with respect to the L^2 metric which is induced from one on v_r^*TX, where

$$v_r(z) = \hat{v}(z, rV_0(z), W_0(z)). \tag{D.3}$$

Let Quad_+ be an open subset of a vector bundle over $\mathfrak{V}$ such that the fiber of Quad_+ at $(z, \mathfrak{v}, \mathfrak{w})$ is the positive definite quadratic form on $T_{u_1(z)}X$. We define its section g as follows. We define $\mathcal{PP} : T_{u_1(z)}X \otimes \Lambda^{0,1}_z \to T_{\hat{v}(z,\mathfrak{v},\mathfrak{w})}X \otimes \Lambda^{0,1}_z$ by

$$\mathcal{PP} = \left(\mathrm{Pal}_{\mathrm{Exp}(u_1(z),\mathfrak{w})}^{\hat{v}(z,\mathfrak{v},\mathfrak{w})}\right)^{(0,1)} \circ \left(\mathrm{Pal}_{u_1(z)}^{\mathrm{Exp}(u_1(z),\mathfrak{w})}\right)^{(0,1)}.$$

Then

$$g(z, \mathfrak{v}, \mathfrak{w})(\vec{v}, \vec{w}) = g_{\hat{v}(z,\mathfrak{v},\mathfrak{w})}(\mathcal{PP}(\vec{v}), \mathcal{PP}(\vec{w}))).$$

Here $\hat{v}$ is as in (D.1) and $g_{\hat{v}(z,\mathfrak{v},\mathfrak{w})}$ is the Riemann metric tensor at $\hat{v}(z, \mathfrak{v}, \mathfrak{w}) \in X$. It is also obvious that g is a smooth section.

We define $\mathbf{e}_i(z, r)$ by induction on i as follows. Suppose $\mathbf{e}_j(z, r)$ is defined for $j < i$. We put

$$\begin{aligned} \mathbf{e}''_i(z, r) =& \mathbf{e}'_i(z, r) \\ & - \sum_{j=1}^{i-1} \int_{z\in\Omega} g(z, rV^0(z), W^0(z))(\mathbf{e}'_i(z, r), \mathbf{e}_j(z, r)) d\,\mathrm{vol} \times \mathbf{e}_j(z, r), \end{aligned} \tag{D.4}$$

and

$$\mathbf{e}_i(z, r) = \frac{\mathbf{e}''_i(z, r)}{\left(\int_{z\in\Omega} g(z, rV^0(z), W^0(z))(\mathbf{e}''_i(z, r), \mathbf{e}''_i(z, r)) d\,\mathrm{vol}\right)^{1/2}}. \tag{D.5}$$

Then $\mathbf{e}_i(z, r)$ $(i = 1, \dots, \dim \mathcal{E}_1)$ is the orthonormal basis we look for.

REMARK D.1. Since the L^2 norm is not defined pointwise, we can not write $\mathbf{e}_i(z, r)$ in a form $\widehat{\mathbf{e}}_i(z, rV^0(z), W^0(z))$ with some smooth map $\widehat{\mathbf{e}}_i(z, \mathfrak{v}, \mathfrak{w})$. In fact (D.4) and (D.5) involve integration. On the other hand, as far as smoothness in the sense of Sobolev space concerns, integration behaves nicely.

Now the linear map

$$\begin{aligned} &((\mathrm{Pal}_{u_1}^{u})^{(0,1)})^{-1} \circ ((\mathrm{Pal}_{u}^{v_r})^{(0,1)})^{-1} \circ \Pi^{\perp}_{\mathcal{E}_1(v_r)} \circ (\mathrm{Pal}_{u}^{v_r})^{(0,1)} \circ (\mathrm{Pal}_{u_1}^{u})^{(0,1)} \\ &: \Gamma(\Omega; u_1^*TX \otimes \Lambda^{0,1}) \to \Gamma(\Omega; u_1^*TX \otimes \Lambda^{0,1}) \end{aligned} \tag{D.6}$$

is written as

$$\mathcal{Q} \mapsto \mathcal{Q} - \sum_{i=1}^{\dim \mathcal{E}_1} \int_{z\in\Omega} g(z, rV^0(z), W^0(z))(\mathcal{Q}(z), \mathbf{e}_i(z, r)) d\,\mathrm{vol} \times \mathbf{e}_i(\cdot, r). \tag{D.7}$$

We write the right hand side of (D.7) by

$$H(r, \mathcal{Q})(\cdot)$$

where $\cdot \in \Omega$. Namely $(r, w) \mapsto H(r, \mathcal{Q})(w)$ is a section of the pullback of $u_1^*TX \otimes \Lambda^{0,1}$ to $[0, \epsilon) \times \Omega$.

By definition of $D_u\mathcal{E}_1$ in (5.7), we have

$$\mathcal{P}^{-1}(D_u\mathcal{E}_1)(\mathcal{P}Q, V)(w) = \left.\frac{d}{dr}\right|_{r=0} H(r, \overline{Q})(w), \tag{D.8}$$

where

$$\overline{Q} = \left((\mathrm{Pal}_{u_1}^{u})^{(0,1)}\right)^{-1}(Q).$$

(Note $Q(z) \in T_{u(z)}X \otimes \Lambda_z^{0,1}$.)

Using (D.8) we can show (C.7) as follows.

Smoothness of $\mathbf{e}'_i$ and (D.2) imply

$$\left\| \frac{d}{dr}\mathbf{e}'_i \right\|_{L^2_m(\Omega)} \le C_{m,(\mathrm{D}.9)}\|V^0\|_{L^2_m}. \tag{D.9}$$

Then using (D.4) and (D.5) we can prove

$$\left\| \frac{d}{dr}\mathbf{e}_i \right\|_{L^2_m(\Omega)} \le C_{m,(\mathrm{D}.10)}\|V^0\|_{L^2_m}. \tag{D.10}$$

Then using also (D.7) and (D.8) we obtain (C.7). □

Estimate of the third term of (C.5). Using (D.2), (D.4), (D.5) in the same way, we obtain

$$\left\| \frac{d^2}{dr^2}\mathbf{e}_i \right\|_{L^2_m(\Omega)} \le C\|V^0\|^2_{L^2_m}.$$

Then using (D.10) and (D.7) also we estimate

$$\left\| \frac{d^2}{dr^2}\Big|_{r=r_0} H(r,\mathcal{Q}) \right\|_{L^2_m} \le C_{m,(\mathrm{D}.11)}\|V^0\|^2_{L^2_m}\|\mathcal{Q}\|_{L^2_m} \tag{D.11}$$

for any $\mathcal{Q} \in L^2_m(\Omega; u_1^*TX \otimes \Lambda^{0,1})$. This gives the required estimate of the third term of (C.5) □

Estimate of (5.82). We put

$$W_0 = \mathrm{E}(u_1, \hat{u}^{\rho}_{1,T,(0)}).$$

Then using (D.10) and (D.8) we obtain

$$\|(D_{\hat{u}^{\rho}_{1,T,(0)}}\mathcal{E}_1)(V,\mathcal{Q})\|_{L^2_m} \le C_{m,(\mathrm{D}.12)}\|V\|_{L^2_m}\|\mathcal{Q}\|_{L^2_m} \tag{D.12}$$

for any $V, \mathcal{Q} \in L^2_m(\Omega; (\hat{u}^{\rho}_{1,T,(0)})^*TX \otimes \Lambda^{01})$.

Using (5.81) also we can estimate

$$\begin{aligned}\Big\|&(D_{\hat{u}^{\rho}_{1,T,(0)}}\mathcal{E}_1)((\mathrm{Pal}_{u_1^{\rho\, b}}^{\hat{u}^{\rho}_{1,T,(0)}})^{0,1}((\mathfrak{se})^{\rho}_{i,T,(\kappa-1)}),\mathcal{Q})\\ &- (D_{\hat{u}^{\rho}_{1,T,(0)}}\mathcal{E}_1)(((\mathrm{Pal}_{u_1^{\rho\, b}}^{\hat{u}^{\rho}_{1,T,(0)}})^{(0,1)}(\mathfrak{e}^{\rho}_{1,T,(0)}),\mathcal{Q})\Big\|_{L^2_m} \le C_{3,m}\frac{e^{-\delta_1 T}}{1-\mu}\|\mathcal{Q}\|_{L^2_m}\end{aligned} \tag{D.13}$$

for L^2_m section $\mathcal{Q}$ of $(\hat{u}^{\rho}_{1,T,(0)})^*TX \otimes \Lambda^{0,1}(\Omega)$.

We put $(\mathrm{Pal}_{u_1^{\rho\, b}}^{\hat{u}^{\rho}_{1,T,(0)}})^{(0,1)}((\mathfrak{se})^{\rho}_{i,T,(\kappa-1)}) = \mathfrak{se}$.

Then using (D.10) to $V_0 = \mathrm{E}(\hat{u}^{\rho}_{1,T,(0)}, \hat{u}^{\rho}_{1,T,(\kappa-1)})$ and integrating along the curve $r \mapsto \mathrm{Exp}(\hat{u}^{\rho}_{1,T,(0)}, rV_0)$ we obtain

$$
\begin{aligned}
&\Big\| (\mathrm{Pal}_{\hat{u}^{\rho}_{1,T,(0)}}^{\hat{u}^{\rho}_{1,T,(\kappa-1)}})^{(0,1)} (D_{\hat{u}^{\rho}_{1,T,(0)}} \mathcal{E}_1)(\mathfrak{se}, \mathcal{Q}) \\
&- (D_{\hat{u}^{\rho}_{1,T,(\kappa-1)}} \mathcal{E}_1)((\mathrm{Pal}_{\hat{u}^{\rho}_{1,T,(0)}}^{\hat{u}^{\rho}_{1,T,(\kappa-1)}})^{(0,1)}(\mathfrak{se}), ((\mathrm{Pal}_{\hat{u}^{\rho}_{1,T,(0)}}^{\hat{u}^{\rho}_{1,T,(\kappa-1)}})^{(0,1)}(\mathcal{Q})) \Big\|_{L^2_m} \\
&\le C_{m,(\mathrm{D}.14)} \|V^0\|_{L^2_m} \|\mathcal{Q}\|_{L^2_m} \le C'_{m,(\mathrm{D}.14)} e^{-\delta_1 T} \|\mathcal{Q}\|_{L^2_m}.
\end{aligned}
\tag{D.14}
$$

Note we use the induction hypothesis (5.69) here.

In the same way as the proof of Lemma 5.20 we can show

$$
\begin{aligned}
&\|((\mathrm{Pal}_{\hat{u}^{\rho}_{1,T,(0)}}^{\hat{u}^{\rho}_{1,T,(\kappa-1)}})^{(0,1)} \circ (\mathrm{Pal}_{u_1^{\rho b}}^{\hat{u}^{\rho}_{1,T,(0)}})^{(0,1)} - (\mathrm{Pal}_{u_1^{\rho b}}^{\hat{u}^{\rho}_{1,T,(\kappa-1)}})^{(0,1)})(\mathfrak{f})\|_{L^2_m} \\
&\le C_{m,(\mathrm{D}.15)} \|\mathrm{E}(\hat{u}^{\rho}_{1,T,(0)}, \hat{u}^{\rho}_{1,T,(\kappa-1)})\|_{L^2_m} \|\mathfrak{f}\|_{L^2_m} \le C'_{m,(\mathrm{D}.15)} e^{-\delta_1 T} \|\mathfrak{f}\|_{L^2_m}
\end{aligned}
\tag{D.15}
$$

and

$$
\begin{aligned}
&\|((\mathrm{Pal}_{\hat{u}^{\rho}_{1,T,(0)}}^{\hat{u}^{\rho}_{1,T,(\kappa-1)}})^{(0,1)} \circ (\mathrm{Pal}_{u_1}^{\hat{u}^{\rho}_{1,T,(0)}})^{(0,1)} - (\mathrm{Pal}_{u_1}^{\hat{u}^{\rho}_{1,T,(\kappa-1)}}))^{(0,1)}(Y)\|_{L^2_m} \\
&\le C_{m,(\mathrm{D}.16)} \|\mathrm{E}(\hat{u}^{\rho}_{1,T,(0)}, \hat{u}^{\rho}_{1,T,(\kappa-1)})\|_{L^2_m} \|Y\|_{L^2_m} \le C'_{m,(\mathrm{D}.16)} e^{-\delta_1 T} \|Y\|_{L^2_m}.
\end{aligned}
\tag{D.16}
$$

Combining (D.12)-(D.16) we obtain the required estimate:

$$
\begin{aligned}
&\Big\| (D_{\hat{u}^{\rho}_{1,T,(\kappa-1)}} \mathcal{E}_1)((\mathrm{Pal}_{u_1^{\rho b}}^{\hat{u}^{\rho}_{1,T,(\kappa-1)}})^{(0,1)}((\mathfrak{se})^{\rho}_{i,T,(\kappa-1)}), ((\mathrm{Pal}_{u_1}^{\hat{u}^{\rho}_{1,T,(\kappa-1)}})^{(0,1)}(Y)) \\
&- (\mathrm{Pal}_{\hat{u}^{\rho}_{1,T,(0)}}^{\hat{u}^{\rho}_{1,T,(\kappa-1)}})^{(0,1)} (D_{\hat{u}^{\rho}_{1,T,(0)}} \mathcal{E}_1)((\mathrm{Pal}_{u_1^{\rho b}}^{\hat{u}^{\rho}_{1,T,(0)}}(\mathfrak{e}^{\rho}_{1,T,(0)}), ((\mathrm{Pal}_{u_1}^{\hat{u}^{\rho}_{1,T,(0)}})^{0,1}(Y)) \Big\|_{L^2_m} \\
&\le C_{m,(\mathrm{D}.17)} e^{-\delta_1 T} \|Y\|_{L^2_m}.
\end{aligned}
\tag{D.17}
$$

□

APPENDIX E

Estimate of the non-linearity of Exponential map

We use the next lemma for the proof of Proposition 6.9 (3).

LEMMA E.1. *Let $\mathscr{S}$ be a 2 dimensional manifold and $u : \mathscr{S} \to X$ a smooth map. Let $v : \mathscr{S} \to X$ be a map of L^2_{m+1} class with $m > 2$. We assume $d(u(z), v(z))$ is smaller than $\iota_X/2$ for all $z \in \mathscr{S}$. We define a map*

$$\mathcal{EP}_v : L^2_{m+1}(\mathscr{S}, u^*TX) \to L^2_{m+1}(\mathscr{S}, u^*TX)$$

by the next formula

$$\mathcal{EP}_v(V)(z) = \mathrm{E}\left(u(z), \mathrm{Exp}(v(z), \mathrm{Pal}_{u(z)}^{v(z)}(V(z)))\right). \tag{E.1}$$

We assume $\|V\|_{L^2_{m+1}}, \|\mathrm{E}(u,v)\|_{L^2_{m+1}} \le 1$. Here $\mathrm{E}(u,v)(z) = \mathrm{E}(u(z), v(z))$. Then

$$\|\mathcal{EP}_v(V) - \mathrm{E}(u,v) - V\|_{L^2_{m+1}} \le C_{m,(\mathrm{E}.2)}\|\mathrm{E}(u,v)\|_{L^2_{m+1}}\|V\|_{L^2_{m+1}}. \tag{E.2}$$

REMARK E.2. In the simplified notation mentioned in Remark 2.2, we have $\mathcal{EP}_v(V) = (v - u) + V$. This way of writing is a bit confusing here since (E.2) implies that this 'equality' holds modulo quadratic order term.

PROOF. Let

$$\Omega = \{(V, W) \in TX \oplus TX \mid |V|, |W| \le \iota'_X/10\}.$$

We define a map $F : \Omega \to TX$ as follows. Let $(V, W) \in T_xX \oplus T_xX$. We put $y = \mathrm{Exp}(x, W)$ and

$$F(V, W) = \mathrm{E}\left(x, \mathrm{Exp}(y, \mathrm{Pal}_x^y(V))\right).$$

This map is obviously smooth. So by the compactness of Ω its C^k norm is uniformly bounded for any k. Moreover we have

$$F(0,0) = 0, \qquad F(V,0) = V, \qquad F(0,W) = W.$$

We observe

$$\mathcal{EP}_v(V)(z) = F(V(z), \mathrm{E}(u(z), v(z))).$$

Therefore the lemma follows from the standard fact that the left composition with smooth map defines a smooth map between L^2_{m+1} spaces. In fact

$$\begin{aligned}\mathcal{EP}_v(V) - \mathrm{E}(u,v) - V &= F(V, \mathrm{E}(u,v)) - F(V,0) - F(0, \mathrm{E}(u,v)) \\ &= \int_0^1 \frac{\partial}{\partial r}(F(V, r\mathrm{E}(u,v)) - F(0, r\mathrm{E}(u,v)))dr \\ &= \int_0^1\int_0^1 \frac{\partial^2}{\partial r \partial t} F(tV, r\mathrm{E}(u,v))drdt.\end{aligned} \tag{E.3}$$

Therefore we can estimate $\mathcal{EP}_v(V) - \mathrm{E}(u,v) - V$ as (E.2), by using

$$\frac{\partial^2}{\partial r \partial t}(F(tV, r\mathrm{E}(u,v)) = \sum_{a,b} V_a \mathrm{E}_b(u,v) G_{a,b}(r,t,tV,r\mathrm{E}(u,v)), \tag{E.4}$$

where V_a, $\mathrm{E}_b(u,v)$ are components of V, $\mathrm{E}(u,v)$, respectively, and $G_{a,b}$ are smooth. □

The following variant of Lemma E.1 can be proved in the same way.

LEMMA E.3. *In the situation of Lemma* E.1 *we assume* $v = v(s)$, $V = V(s)$ *are families of* C^ℓ *class parameterized by* $s \in \mathcal{S}$ *for some parameter space* $\mathcal{S} \subset \mathbb{R}^N$. *We assume* $\left\|\frac{d^{n'}}{ds^{n'}} V\right\|_{L^2_{m+1}}$, $\left\|\frac{d^{n'}}{ds^{n'}} \mathrm{E}(u,v)\right\|_{L^2_{m+1}} \le 1$ *for* $n' \le n$. *Then*

$$\begin{aligned}&\left\|\frac{d^n}{ds^n}\left(\mathcal{EP}_{v(s)}(V(s)) - \mathrm{E}(u,v(s)) - V(s)\right)\right\|_{L^2_{m+1}} \\ &\le C_{m,(\mathrm{E}.5)} \sum_{k_1+k_2 \le n} \left\|\frac{d^{k_1}}{ds^{k_1}} V(s)\right\|_{L^2_{m+1}} \left\|\frac{d^{k_2}}{ds^{k_2}} \mathrm{E}(u,v(s))\right\|_{L^2_{m+1}}.\end{aligned} \tag{E.5}$$

PROOF. The lemma follows easily by taking s differentiation of (E.3) and (E.4). □

We can prove Proposition 5.23 (3) by using Lemma E.3. Actually Proposition 5.23 (3) follows from (6.10) in the same way as the proof of Lemma 6.12, that is, by integration over T.

PROOF OF PROPOSITION 6.9 (3). Using the simplified notation as in Remark 2.2 we have

$$u^\rho_{T,(\kappa)} \text{``}=\text{''} u^{\rho_1} + \sum_{j=0}^{\kappa} V^\rho_{T,1,(j)}$$

on K_1. Therefore

$$\begin{aligned}\|u^\rho_{T,(\kappa)} - u^\rho_1\|_{L^2_{m+1}} &\text{``}\le\text{''} \sum_{j=0}^{\kappa} \|V^\rho_{T,1,(j)}\|_{L^2_{m+1}} \\ &\text{``}\le\text{''} C_{5,m} \sum_{j=0}^{\kappa} \mu^{j-1} e^{-\delta_1 T} \le C_{5,m}\mu^{-1}\frac{1-\mu^{\kappa+1}}{1-\mu} e^{-\delta_1 T}.\end{aligned}$$

This would prove (6.10) on K_1. However here the above notation $+$ is rather imprecise and we need to use exponential map and parallel transport. In other words we need to estimate the effect of the nonlinearity of the exponential map to work out the above inequality. We use Lemma E.3 for this purpose. The above inequality suggests that $C_{6,m}$ depends on $C_{5,m}$.

In the proof of Proposition 6.9 (3), we need to decide $C_{6,m}$ and $C_{7,m}$ depending on $C_{5,m}$, $C_{8,m}$, $C_{9,m}$. So the dependence of the constants appearing during the proof on $C_{5,m}$, $C_{6,m}$, $C_{7,m}$, $C_{8,m}$, $C_{9,m}$ need to be carefully examined.

Hereafter during the proof of Proposition 6.9 (3) we use the notation of the constants $C_{m,*}$ and $D_{m,*}$ as follows. (Here $*$ is the number of the formula where the constants appear.) $C_{m,*}$ is a constant depending on all of $C_{5,m}$, $C_{6,m}$, $C_{7,m}$, $C_{8,m}$, $C_{9,m}$ and on m. $D_{m,*}$ is a constant depending $C_{5,m}$, $C_{8,m}$, $C_{9,m}$, m *but is* *independent of* $C_{6,m}$ *and* $C_{7,m}$. $T_{m,*}$ is a constant depending on all of $C_{5,m}$, $C_{6,m}$, $C_{7,m}$, $C_{8,m}$, $C_{9,m}$ and on m.

We now start the proof. During the proof we assume $m-2 \geq n, \ell \geq 0$. .

On K_1 we have $u^{\rho}_{T,(\kappa)} = \operatorname{Exp}(u^{\rho}_{T,(\kappa-1)}, V^{\rho}_{T,1,(\kappa)})$. Therefore by (6.9) for $\leq \kappa$, (6.10) for $\leq \kappa - 1$, and Lemma E.3 we have

$$\begin{aligned}
&\left\| \nabla^n_{\rho} \frac{\partial^{\ell}}{\partial T^{\ell}} \mathscr{P}_{1,\rho_1} \mathrm{E}(u^{\rho}_1, u^{\rho}_{T,(\kappa)}) \right\|_{L^2_{m+1-\ell,\delta}(K_1)} \\
&< \left\| \nabla^n_{\rho} \frac{\partial^{\ell}}{\partial T^{\ell}} \mathscr{P}_{1,\rho_1} \mathrm{E}(u^{\rho}_1, u^{\rho}_{T,(\kappa-1)}) \right\|_{L^2_{m+1-\ell,\delta}(K_1)} \\
&\quad + C_{5,m} \mu^{\kappa-1} e^{-\delta_1 T} + C_{m,(\mathrm{E}.6)} \mu^{\kappa-1} e^{-2\delta_1 T}.
\end{aligned} \tag{E.6}$$

Taking $T_{m,(\mathrm{E}.7)}$ so that $C_{m,(\mathrm{E}.6)} e^{-\delta_1 T_{m,(\mathrm{E}.7)}} \leq 1$ we have

$$\begin{aligned}
&\text{LHS of (E.6)} \\
&\leq \left\| \nabla^n_{\rho} \frac{\partial^{\ell}}{\partial T^{\ell}} \mathscr{P}_{1,\rho_i} \mathrm{E}(u^{\rho}_1, u^{\rho}_{T,(\kappa-1)}) \right\|_{L^2_{m+1-\ell,\delta}(K_1)} + D_{m,(\mathrm{E}.7)} \mu^{\kappa-1} e^{-\delta_1 T}
\end{aligned} \tag{E.7}$$

for $T \geq T_{m,(\mathrm{E}.7)}$. Here $\mathscr{P}_{1,\rho_i}$ is defined in (6.8).

We next study at the neck region. We consider $m-2 \geq n, \ell \geq 0$. (The case $\ell = 0$ is included.) We put

$$U^{\rho}_{T,i,(\kappa)}(\tau, t) = V^{\rho}_{T,i,(\kappa)}(\tau, t) - (\Delta p^{\rho}_{T,(\kappa)})^{\mathrm{Pal}} \tag{E.8}$$

and

$$U(\tau', t) = U^{\rho}_{T,1,(\kappa)}(\tau', t) + \chi(\tau' - 4T) U^{\rho}_{T,2,(\kappa)}(\tau' - 10T, t). \tag{E.9}$$

On $(\tau', t) \in [0, 5T+1]_{\tau'} \times [0,1]$, we have

$$\begin{aligned}
u^{\rho}_{T,(\kappa)}(\tau', t) &= \operatorname{Exp}\Big(u^{\rho}_{T,(\kappa-1)}(\tau', t), V^{\rho}_{T,1,(\kappa)}(\tau', t) \\
&\qquad + \chi(\tau' - 4T)(U^{\rho}_{T,2,(\kappa)}(\tau' - 10T, t)\Big) \\
&= \operatorname{Exp}\Big(u^{\rho}_{T,(\kappa-1)}(\tau', t), U(\tau', t) + (\Delta p^{\rho}_{T,(\kappa)})^{\mathrm{Pal}}\Big).
\end{aligned} \tag{E.10}$$

Suppose $[S', S'+1]_{\tau'} \subseteq [0, 5T+1]_{\tau'} \times [0,1]$. We put

$$\mathfrak{P}_i = \left(\mathrm{Pal}_{u_i}^{u^{\rho}_{T,(\kappa-1)}}\right)^{-1} = \mathrm{Pal}^{u_i}_{u^{\rho}_{T,(\kappa-1)}}$$

and

$$\Sigma(S') = [S', S'+1]_{\tau'} \times [0,1] \subset [0, 5T+1]_{\tau'} \times [0,1].$$

When we use the τ' (resp. τ'') coordinate we write $\Sigma(S')_{\tau'}$ (resp. $\Sigma(S')_{\tau''}$). Using (6.36) we can estimate

$$\begin{aligned}
&\left\| \nabla^n_{\rho} \frac{\partial^{\ell}}{\partial T^{\ell}} \mathfrak{P}_1 U(\tau', t) \right\|_{L^2_{m+1-\ell}(\Sigma(S')_{\tau'})} \\
&\leq \sum_{l=1}^{\ell} \sum_{n'=0}^{n} D_{m,(\mathrm{E}.11)} \left(\left\| \left(\nabla^{n'}_{\rho} \frac{\partial^{l}}{\partial T^{l}} \right) (\mathfrak{P}_1 U^{\rho}_{T,1,(\kappa)}(\tau', t)) \right\|_{L^2_{m-l+1}(\Sigma(S')_{\tau'})} \right. \\
&\qquad \left. + \left\| \left(\nabla^{n'}_{\rho} \frac{\partial^{l}}{\partial T^{l}} \right) (\mathfrak{P}_1 U^{\rho}_{T,2,(\kappa)}(\tau'', t)) \right\|_{L^2_{m+1}(\Sigma(S')_{\tau''})} \right)
\end{aligned} \tag{E.11}$$

We remark that the second term of (E.11) drops if $S' \le T$. (This is because $\chi(\tau' - 4T) = 0$ on $\Sigma(S')_{\tau'}$ in that case.)

By (6.9) we have

$$\sum_{S'=-5T}^{0} \left\| \left(\nabla_\rho^{n'} \frac{\partial^l}{\partial T^l}\right) (\mathfrak{P}_i U^\rho_{T,i,(\kappa)}(\tau', t)) \right\|_{L^2_{m-l+1,\delta}(\Sigma(S')_{\tau'})} \le C_{5,m} \mu^{\kappa-1} e^{-\delta_1 T}, \tag{E.12}$$

for $i = 1, 2$. In case $i = 2$ we need to replace $\mathfrak{P}_2$ by $\mathfrak{P}_1$ to apply (E.12) to estimate (E.11).

SUBLEMMA E.4. *We have*

$$\begin{aligned} &\left\| \left(\nabla_\rho^{n'} \frac{\partial^l}{\partial T^l}\right) (\mathfrak{P}_1 U^\rho_{T,2,(\kappa)}(\tau'', t)) \right\|_{L^2_{m+1-l}(\Sigma(S')_{\tau''})} \\ &\le D_{m,(\mathrm{E}.13)} \sum_{l' \le l} \sum_{n'' \le n'} \left\| \left(\nabla_\rho^{n''} \frac{\partial^{l'}}{\partial T^{l'}}\right) (\mathfrak{P}_2 U^\rho_{T,2,(\kappa)}(\tau'', t)) \right\|_{L^2_{m+1-l'}(\Sigma(S')_{\tau''})} \end{aligned} \tag{E.13}$$

if $T > T_{m,(\mathrm{E}.13)}$ *and* $S' \ge T$.

Postponing the proof of the sublemma we continue the proof. Note the ratio between the weights $e_{1,\delta}$ and $e_{T,\delta}$ are bounded on $\tau' \in [0, 5T]_{\tau'} \times [0, 1]$ and the weight $e_{2,\delta}$ is larger than $e_{T,\delta}$. Therefore taking the weighted sum of the square of (E.11) for $S' = 0, \dots, [5T], 5T$ (where $[5T]$ is the largest integer with $\le 5T$) and using (E.12), (E.13) we obtain

$$\left\| \nabla_\rho^n \frac{\partial^\ell}{\partial T^\ell} \mathfrak{P}_1 U(\tau', t) \right\|_{L^2_{m+1-\ell,\delta}(K_1^{5T+1} \subset \Sigma_T)} \le D_{m,(\mathrm{E}.14)} \mu^{\kappa-1} e^{-\delta_1 T}. \tag{E.14}$$

We put $q = u_1(0, 1/2)$ and

$$\mathfrak{P}_1' = \mathrm{Pal}_q^{u_1} \circ \mathrm{Pal}^q_{u^\rho_{T,(\kappa-1)}}$$

SUBLEMMA E.5. *We have*

$$\begin{aligned} &\Big\| \nabla_\rho^n \frac{\partial^\ell}{\partial T^\ell} \Big(\mathfrak{P}_1((\Delta p^\rho_{T,(\kappa)})^{\mathrm{Pal}}) \\ &\qquad - \mathfrak{P}_1'((\Delta p^\rho_{T,(\kappa)})^{\mathrm{Pal}}) \Big) \Big\|_{L^2_{m+1-\ell,\delta}([0,5T+1]_{\tau'} \times [0,1])} \\ &\le D_{m,(\mathrm{E}.15)} \mu^{\kappa-1} e^{-\delta_1 T}. \end{aligned} \tag{E.15}$$

Here we regard $(\Delta p^\rho_{T,(\kappa)})^{\mathrm{Pal}}$ as a section on $(u^\rho_{T,(\kappa-1)})^* TX$ (as we do in (E.10)). So (E.15) is

$$\begin{aligned} &\Big\| \nabla_\rho^n \frac{\partial^\ell}{\partial T^\ell} \Big((\mathrm{Pal}^{u_1}_{u^\rho_{T,(\kappa-1)}} \circ \mathrm{Pal}^{u^\rho_{T,(\kappa-1)}}_{p^\rho_{T,(\kappa-1)}})(\Delta p^\rho_{T,(\kappa)}) \\ &- (\mathrm{Pal}_q^{u_1} \circ \mathrm{Pal}^q_{u^\rho_{T,(\kappa-1)}} \circ \mathrm{Pal}^{u^\rho_{T,(\kappa-1)}}_{p^\rho_{T,(\kappa-1)}})(\Delta p^\rho_{T,(\kappa)}) \Big) \Big\|_{L^2_{m+1-\ell}([0,5T+1]_{\tau'} \times [0,1])} \\ &\le D_{m,(\mathrm{E}.15)} \mu^{\kappa-1} e^{-\delta_1 T}. \end{aligned} \tag{E.16}$$

Postponing the proof of the sublemma we continue the proof.

Lemma E.3 and (E.10), (E.14), (E.15), imply

$$\begin{aligned}(\text{E.17})\quad &\left\Vert \nabla_\rho^n \frac{\partial^\ell}{\partial T^\ell}\Big(\mathfrak{P}_1 \mathrm{E}(u^\rho_{T,(\kappa-1)}, u^\rho_{T,(\kappa)}) \right.\\ &\qquad \left. - \mathfrak{P}_1'\Big((\Delta p^\rho_{T,(\kappa)})^{\mathrm{Pal}}\Big)\Big)\right\Vert_{L^2_{m+1-\ell,\delta}([0,5T+1]_{\tau'}\times[0,1]\subset\Sigma_T)}\\ &\le D_{m,(\text{E.17})}\mu^{\kappa-1}e^{-\delta_1 T}.\end{aligned}$$

Using also Lemma E.3 we can show the next sublemma.

SUBLEMMA E.6. *The next inequality holds.*

$$\begin{aligned}(\text{E.18})\quad &\left\Vert \nabla_\rho^n \frac{\partial^\ell}{\partial T^\ell}\mathscr{P}_{1,\rho_1}\mathrm{E}(u_1^\rho, u^\rho_{T,(\kappa)})\right\Vert_{W^2_{m+1-\ell,\delta}([0,5T+1]_{\tau'}\times[0,1]\subset\Sigma_T)}\\ &\le \left\Vert \nabla_\rho^n \frac{\partial^\ell}{\partial T^\ell}\mathscr{P}_{1,\rho_1}\mathrm{E}(u_1^\rho, u^\rho_{T,(\kappa-1)})\right\Vert_{W^2_{m+1-\ell,\delta}([0,5T+1]_{\tau'}\times[0,1]\subset\Sigma_T)}\\ &\quad + D_{m,(\text{E.18})}\mu^{\kappa-1}e^{-\delta_1 T}.\end{aligned}$$

Note the term corresponding to

$$\mathfrak{P}_1'((\Delta p^\rho_{T,(\kappa)})^{\mathrm{Pal}}) = (\mathrm{Pal}_q^{u_1}\circ \mathrm{Pal}^q_{u^\rho_{T,(\kappa-1)}})((\Delta p^\rho_{T,(\kappa)})^{\mathrm{Pal}})$$

disappears in (E.18) since we take $W^2_{m+1-\ell',\delta}$ norm in place of $L^2_{m+1-\ell',\delta}$ norm. See (4.9) and we recall $q = u_1(0,1/2)$. Postponing the detail of the proof of the sublemma we continue the proof.

We choose $C_{6,m}$ such that

$$D_{m,(\text{E.18})} \le C_{6,m}(1-\mu)/10, \quad D_{m,(\text{E.7})} \le C_{6,m}(1-\mu)/10.$$

Then (E.7), (E.18) and the $(\kappa-1)$-th case of (6.11) implies that for $\ell > 0$ we have

$$\begin{aligned}(\text{E.19})\quad &\left\Vert \nabla_\rho^n \frac{\partial^\ell}{\partial T^\ell}\mathscr{P}_{1,\rho_1}\mathrm{E}(u_1^\rho, u^\rho_{T,(\kappa)})\right\Vert_{W^2_{m+1-\ell,\delta}(K_1^{5T+1})}\\ &\le \left\Vert \nabla_\rho^n \frac{\partial^\ell}{\partial T^\ell}\mathscr{P}_{1,\rho_1}\mathrm{E}(u_1^\rho, u^\rho_{T,(\kappa-1)})\right\Vert_{W^2_{m+1-\ell,\delta}(K_1^{5T+1})}\\ &\quad + D_{m,(\text{E.18})}\mu^{\kappa-1}e^{-\delta_1 T} + D_{m,(\text{E.7})}\mu^{\kappa-1}e^{-\delta_1 T}\\ &\le C_{6,m}(2-\mu^{\kappa-1})e^{-\delta_1 T} + \frac{C_{6,m}(\mu^{\kappa-1}-\mu^\kappa)e^{-\delta_1 T}}{5}\\ &\le C_{6,m}(2-\mu^{\kappa})e^{-\delta_1 T}.\end{aligned}$$

We thus proved the κ-th case of (6.10).

We next prove (6.11). We consider $m-2 \ge n, \ell \ge 0$. (The case $\ell = 0$ is included.) We denote

$$d_{\ell,n}(u,v) = \sum_{\substack{n'\le n\\ \ell'\le \ell}} \left\Vert \nabla_\rho^{n'} \frac{\partial^{\ell'}}{\partial T^{\ell'}}\mathrm{Pal}_u^{u_1}\mathrm{E}(u,v)\right\Vert_{L^2_{m+1-\ell'}(K_1^{9T}\setminus K_1^T)}.$$

Here u, v are (T,ρ) dependent family of maps on Σ_1. (We assume $d(u_1(z), u(z)) \le \iota_X$ etc. so that $\mathrm{Pal}^u_{u_1}$ etc. is defined.) We remark that we take $L^2_{m+1-\ell'}$ norm without weight in the right hand side.

Note $d_{\ell,n}$ is not a metric. In particular it may not satisfy triangle inequality. However Lemma E.3 implies

$$d_{\ell,n}(v_1,v_3) \le d_{\ell,n}(v_1,v_2) + d_{\ell,n}(v_2,v_3) + C_{m,(\mathrm{E}.20)} d_{\ell,n}(v_1,v_2) d_{\ell,n}(v_2,v_3) \tag{E.20}$$

if

$$\left\| \nabla_\rho^{n'} \frac{\partial^{\ell'}}{\partial T^{\ell'}} \mathrm{E}(u_1, v_i) \right\|_{L^2_{m+1-\ell'}(K_1^{9T} \setminus K_1^T)} \le 1$$

for $i = 1,2,3$, $n' \le n$, $\ell' \le \ell$.

Note (E.17) holds when we replace $\mathfrak{P}'_1$ there by $\mathfrak{P}_1$. It then implies

$$d_{\ell,n}(u^\rho_{T,(\kappa)}, \mathrm{Exp}(u^\rho_{T,(\kappa-1)}, (\Delta p^\rho_{T,(\kappa)})^{\mathrm{Pal}})) \le D_{m,(\mathrm{E}.21)} \mu^{\kappa-1} e^{-\delta_1 T}. \tag{E.21}$$

We put

$$\square_{\kappa-1} = \mathrm{E}(p_0^\rho, u^\rho_{T,(\kappa-1)}).$$

The induction hypothesis, which is $\kappa-1$ case of (6.11), implies

$$\left\| \nabla_\rho^{n} \frac{\partial^{\ell}}{\partial T^{\ell}} \mathrm{Pal}_{p_0^\rho}^{p_0}(\square_{\kappa-1}) \right\|_{L^2_{m+1-\ell}(K_1^{9T} \setminus K_1^T)} \le C_{7,m}(2-\mu^{\kappa-1}) e^{-\delta_1 T}. \tag{E.22}$$

Note (6.11) is actually (6.17).

Then Lemma E.3 implies

$$\begin{aligned} &d_{\ell,n}(\mathrm{Exp}(u^\rho_{T,(\kappa-1)}, (\Delta p^\rho_{T,(\kappa)})^{\mathrm{Pal}}), \mathrm{Exp}(p_0^\rho, \square_{\kappa-1} + \mathrm{Pal}_{p^\rho_{T,(\kappa-1)}}^{p_0^\rho}(\Delta p^\rho_{T,(\kappa)})) \\ &\le C_{m,(\mathrm{E}.23)} \mu^{\kappa-1} e^{-2\delta_1 T}. \end{aligned} \tag{E.23}$$

On the other hand, when we regard $(\Delta p^\rho_{T,(\kappa)})^{\mathrm{Pal}}$ as an element of $T_{p_0^\rho} X$ we have

$$\begin{aligned} &\left\| \nabla_\rho^{n} \frac{\partial^{\ell}}{\partial T^{\ell}} \mathrm{Pal}_{p_0^\rho}^{p_0}(\mathrm{Pal}_{p^\rho_{T,(\kappa-1)}}^{p_0^\rho}(\Delta p^\rho_{T,(\kappa)})) \right\|_{L^2_{m+1-\ell}(K_1^{9T} \setminus K_1^T)} \\ &\le D_{m,(\mathrm{E}.24)} \mu^{\kappa-1} e^{-\delta_1 T}. \end{aligned} \tag{E.24}$$

This follows from (6.9), Lemma B.1 and

$$\left\| \nabla_\rho^{n} \frac{\partial^{\ell}}{\partial T^{\ell}} \mathrm{Pal}_{p_0^\rho}^{p_0}(\mathrm{E}(p_0^\rho, p^\rho_{T,(\kappa)})) \right\|_{L^2_{m+1-\ell}(K_1^{9T} \setminus K_1^T)} \le C_{m,(\mathrm{E}.25)} \mu^{\kappa-1} e^{-\delta_1 T}. \tag{E.25}$$

(E.21), (E.23) and (E.20) imply

$$d_{\ell,n}(u^\rho_{T,(\kappa)}, \mathrm{Exp}(p_0^\rho, \square_{\kappa-1} + \mathrm{Pal}_{p^\rho_{T,(\kappa-1)}}^{p_0^\rho}(\Delta p^\rho_{T,(\kappa)})) \le D_{m,(\mathrm{E}.26)} \mu^{\kappa-1} e^{-\delta_1 T} \tag{E.26}$$

if $T > T_{m,(\mathrm{E}.26)}$. We use the fact that the exponent of e in (E.23) is $-2\delta_1 T$, to change $C_{m,(\mathrm{E}.23)}$ there to $D_{m,(\mathrm{E}.26)}$. The third term of (E.20) is estimated by $C\mu^{2(\kappa-1)} e^{-2\delta_1 T}$ in our case. So it is estimated by $D\mu^{\kappa-1} e^{-\delta_1 T}$ if T is sufficiently large.

Using Lemma E.3, (E.24), (E.22), (E.26) imply

$$\begin{aligned} &\left\| \nabla_\rho^{n} \frac{\partial^{\ell}}{\partial T^{\ell}} \mathrm{Pal}_{p_0^\rho}^{p_0}(\mathrm{E}(p_0^\rho, u^\rho_{T,(\kappa)})) \right\|_{L^2_{m+1-\ell}(K_1^{9T} \setminus K_1^T)} \\ &\le C_{7,m}(2-\mu^{\kappa-1}) e^{-\delta_1 T} + D_{m,(\mathrm{E}.27)} \mu^{\kappa-1} e^{-\delta_1 T}. \end{aligned} \tag{E.27}$$

We choose $C_{7,m}$ such that

$$C_{7,m}(1-\mu) > D_{m,(\mathrm{E}.27)}.$$

Then

$$\left\|\nabla_\rho^n \frac{\partial^\ell}{\partial T^\ell}\mathrm{Pal}_{p_0^\rho}^{p_0}\mathrm{E}(p_0^\rho, u_{T,(\kappa)}^\rho)\right\|_{L^2_{m+1-\ell}(K_1^{9T}\setminus K_1^T)} \le C_{7,m}(2-\mu^\kappa)e^{-\delta_1 T}. \tag{E.28}$$

REMARK E.7. We remark that when we consider the domain $K_1^{9T}\setminus K_1$ in place of $K_1^{9T}\setminus K_1^T$ and $\ell = 0$ then (E.28) fails in the first step, that is, $\kappa = 0$.

To complete the proof of Proposition 6.9 (3) it remains to prove sublemmata. □

PROOF OF SUBLEMMA E.4. We take and fix ρ_0' and prove this estimate under the assumption that $d(\rho, \rho_0') \le e^{-\delta_1 T}$. As usual we use the τ' (resp τ'') coordinate in the definition of the L^2_{m+1} norm for $i = 1$ (resp. $i = 2$). We consider maps

$$\mathfrak{P}_i'' = \mathrm{Pal}_{p^{\rho_0'}}^{u_i} : \Gamma(\Sigma(S'); T_{p^{\rho_0'}}X) \to \Gamma(\Sigma(S'); u_i^*TX).$$

Let W be a (T,ρ) parameterized family of sections of $\Gamma(\Sigma(S'); T_{p^{\rho_0'}}X)$. By using (T,ρ) independence of $\mathfrak{P}_i''$ we can prove

$$\begin{aligned} D_{m,(\mathrm{E}.29)}^{-1} &\left\|\nabla_\rho^n \frac{\partial^l}{\partial T^l}\mathfrak{P}_i''(W)\right\|_{L^2_{m+1-l}(\Sigma(S'))} \\ &\le \left\|\nabla_\rho^n \frac{\partial^l}{\partial T^l} W\right\|_{L^2_{m+1-l}(\Sigma(S'))} \\ &\le D_{m,(\mathrm{E}.29)}\left\|\nabla_\rho^n \frac{\partial^l}{\partial T^l}\mathfrak{P}_i''(W)\right\|_{L^2_{m+1-l}(\Sigma(S'))} \end{aligned} \tag{E.29}$$

for $i = 1, 2$.

Since $S' \ge T$ we have

$$\|\mathrm{E}(p_0^{\rho_0'}, u_{T,(\kappa-1)}^\rho)\|_{L^2_{m+1}([S',S'+1]_{\tau'}\times[0,1])} \le C_{m,(\mathrm{E}.30)}e^{-\delta_1 T}. \tag{E.30}$$

Here $C_{m,(\mathrm{E}.30)}$ depends on $C_{7,m}$.

We put:

$$\mathfrak{P}''' = \mathrm{Pal}_{p_0^{\rho_0'}}^{u_{T,(\kappa-1)}^\rho} : \Gamma(\Sigma(S'); T_{p_0^{\rho_0'}}X) \to \Gamma(\Sigma(S'); (u_{T,(\kappa-1)}^\rho)^*TX).$$

If $\hat W$ is a (T,ρ) independent section of u_2^*TX on $\Sigma(S')$, Lemma B.1 implies

$$\begin{aligned} &\|\nabla_\rho^n \frac{\partial^l}{\partial T^l}(\mathfrak{P}_1''\circ(\mathfrak{P}_2'')^{-1} - \mathfrak{P}_1\circ(\mathfrak{P}_2)^{-1})(\hat W)\|_{L^2_{m+1-l}(\Sigma(S'))} \\ &\le \|\nabla_\rho^n \frac{\partial^l}{\partial T^l}(\mathfrak{P}_1''\circ(\mathfrak{P}_2'')^{-1} - \mathfrak{P}_1\circ\mathfrak{P}'''\circ(\mathfrak{P}_2'')^{-1})(\hat W)\|_{L^2_{m+1-l}(\Sigma(S'))} \\ &\quad + \|\nabla_\rho^n \frac{\partial^l}{\partial T^l}(\mathfrak{P}_1\circ\mathfrak{P}'''\circ(\mathfrak{P}_2'')^{-1} - (\mathfrak{P}_1\circ(\mathfrak{P}_2)^{-1})(\hat W)\|_{L^2_{m+1-l}(\Sigma(S'))} \\ &\le C_{m,(\mathrm{E}.31)}e^{-\delta_1 T}\|\hat W\|_{L^2_{m+1-l}(\Sigma(S'))} \end{aligned} \tag{E.31}$$

See Figure E.1.

We put

$$\hat W(T,\rho) = \mathfrak{P}_2(U_{T,2,(\kappa)}^\rho).$$

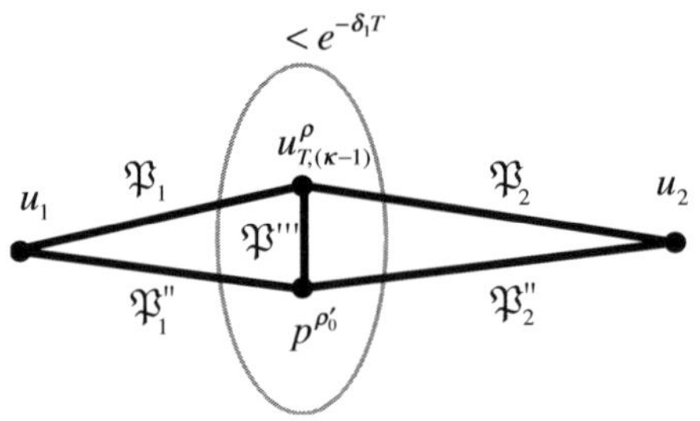

FIGURE E.1. $\mathfrak{P}'''$, $\mathfrak{P}''_i$ and $\mathfrak{P}_i$

Then using (E.12), (E.31) we can derive

$$
\begin{aligned}
&\left\|\nabla_\rho^{n'}\frac{\partial^l}{\partial T^l}\mathfrak{P}_1(U^\rho_{T,2,(\kappa)}) - \mathfrak{P}''_1(\mathfrak{P}''_2)^{-1}(\hat W(T,\rho)))\right\|_{L^2_{m+1-l}(\Sigma(S')_{\tau''})}\\
&= \left\|\nabla_\rho^{n'}\frac{\partial^l}{\partial T^l}(\mathfrak{P}_1\circ\mathfrak{P}_2^{-1} - \mathfrak{P}''_1\circ(\mathfrak{P}''_2)^{-1})(\hat W(T,\rho))\right\|_{L^2_{m+1-l}(\Sigma(S')_{\tau''})}\\
&\le \sum_{n''\le n'}\sum_{l'\le l} C_{m,(\mathrm{E}.32)}e^{-\delta_1 T}\left\|\nabla_\rho^{n''}\frac{\partial^{l'}}{\partial T^{l'}}\hat W(T,\rho)\right\|_{L^2_{m+1-l'}(\Sigma(S')_{\tau''})}.
\end{aligned}
\tag{E.32}
$$

Therefore using (E.29) also we have

$$
\begin{aligned}
&\left\|\left(\nabla_\rho^{n'}\frac{\partial^l}{\partial T^l}\right)(\mathfrak{P}_1 U^\rho_{T,2,(\kappa)}(\tau'',t))\right\|_{L^2_{m+1-l}(\Sigma(S')_{\tau''})}\\
&\le (D_{m,(\mathrm{E}.33)} + C_{m,(\mathrm{E}.33)}e^{-\delta_1 T})\\
&\qquad \sum_{n''\le n'}\sum_{l'\le l}\left\|\nabla_\rho^{n''}\frac{\partial^{l'}}{\partial T^{l'}}\hat W(T,\rho)\right\|_{L^2_{m+1-l'}(\Sigma(S')_{\tau''})}.
\end{aligned}
\tag{E.33}
$$

We take $T_{m,(\mathrm{E}.13)}$ such that $e^{-\delta_1 T}C_{m,(\mathrm{E}.33)} \le 1$ if $T > T_{m,(\mathrm{E}.13)}$. The sublemma is proved. □

PROOF OF SUBLEMMA E.5. The proof is similar to the proof of Sublemma E.4. We take and fix ρ'_0 and prove this estimate under the assumption that $d(\rho,\rho'_0) \le e^{-\delta_1 T}$.

We observe

$$
\begin{aligned}
\left\|\nabla_\rho^n\frac{\partial^\ell}{\partial T^\ell}\mathrm{E}(u_1^{\rho'_0}, u^\rho_{T,(\kappa-1)})\right\|_{L^2_{m+1-\ell}(\Sigma(S')_{\tau'})} &\le C_{m,(\mathrm{E}.34)}e^{-\delta_1 T},\\
\left|\nabla_\rho^n\frac{\partial^\ell}{\partial T^\ell}\mathrm{E}(p_0^{\rho'_0}, p^\rho_{T,(\kappa-1)})\right| &\le C'_{m,(\mathrm{E}.34)}e^{-\delta_1 T}.
\end{aligned}
\tag{E.34}
$$

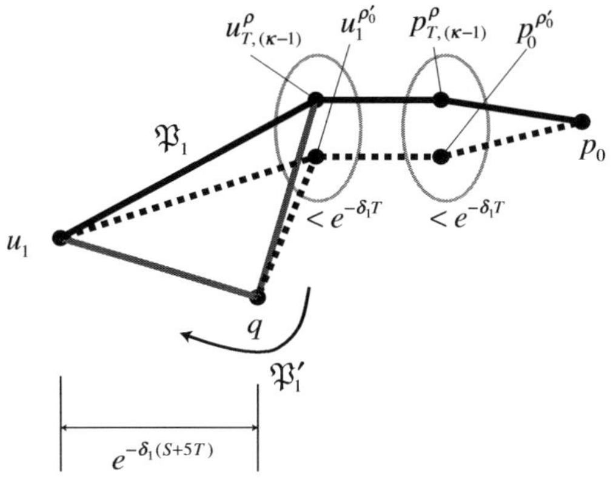

FIGURE E.2. Parallel transportations in the proof of Sublemma E.5

Therefore using an obvious variant of Lemma B.1 we can prove

$$\begin{aligned}&\Big\|\nabla_\rho^n\frac{\partial^\ell}{\partial T^\ell}\Big(\mathfrak{P}_1\circ \mathrm{Pal}^{u^{\rho}_{T,(\kappa-1)}}_{p^{\rho}_{T,(\kappa-1)}}\circ \mathrm{Pal}^{p^{\rho}_{T,(\kappa-1)}}_{p_0}\\ &\qquad -\mathrm{Pal}^{u_1}_{u_1^{\rho_0'}}\circ \mathrm{Pal}^{u_1^{\rho_0'}}_{p_0^{\rho_0'}}\circ \mathrm{Pal}^{p_0^{\rho_0'}}_{p_0}\Big)(W)\Big\|_{L^2_{m+1-\ell}(\Sigma(S')_{\tau'})}\\ &\le C_{m,(\mathrm{E}.35)}e^{-\delta_1 T}\|W\|\end{aligned} \tag{E.35}$$

and

$$\begin{aligned}&\Big\|\nabla_\rho^n\frac{\partial^\ell}{\partial T^\ell}\Big(\mathfrak{P}_1'\circ \mathrm{Pal}^{u^{\rho}_{T,(\kappa-1)}}_{p^{\rho}_{T,(\kappa-1)}}\circ \mathrm{Pal}^{p^{\rho}_{T,(\kappa-1)}}_{p_0}\\ &\qquad -\mathrm{Pal}^{u_1}_{q}\circ \mathrm{Pal}^{q}_{u_1^{\rho_0'}}\circ \mathrm{Pal}^{u_1^{\rho_0'}}_{p_0^{\rho_0'}}\circ \mathrm{Pal}^{p_0^{\rho_0'}}_{p_0}\Big)(W)\Big\|_{L^2_{m+1-\ell}(\Sigma(S')_{\tau'})}\\ &\le C_{m,(\mathrm{E}.36)}e^{-\delta_1 T}\|W\|\end{aligned} \tag{E.36}$$

for $W \in T_{p_0}X$. See Figure E.2.

Note (6.9) and (6.17) imply

$$\Big\|\nabla_\rho^n\frac{\partial^\ell}{\partial T^\ell}\Big(\mathrm{Pal}^{p_0}_{p^{\rho}_{T,(\kappa-1)}}(\Delta p^{\rho}_{T,(\kappa)})\Big)\Big\|_{L^2_{m+1-\ell}(\Sigma(S')_{\tau'})} \le C_{m,(\mathrm{E}.37)}\mu^{\kappa-1}e^{-\delta_1 T}. \tag{E.37}$$

On the other hand a similar formula as (E.29) holds with $\mathfrak{P}_i''$ replaced by $\mathrm{Pal}^{p_0^{\rho_0'}}_{p_0}$ or by $\mathrm{Pal}^{u_i^{\rho_0'}}_{p_0^{\rho_0'}}$. Therefore we obtain

$$\begin{aligned}&\Big\|\nabla_\rho^n\frac{\partial^\ell}{\partial T^\ell}\Big((\mathrm{Pal}^{u_1^{\rho_0'}}_{p_0^{\rho_0'}}\circ \mathrm{Pal}^{p_0^{\rho_0'}}_{p_0}\circ \mathrm{Pal}^{p_0}_{p^{\rho}_{T,(\kappa-1)}})(\Delta p^{\rho}_{T,(\kappa)})\Big)\Big\|_{L^2_{m+1-\ell}(\Sigma(S'))}\\ &\le C_{m,(\mathrm{E}.38)}\mu^{\kappa-1}e^{-\delta_1 T}.\end{aligned} \tag{E.38}$$

By Lemma 2.5 and Condition 3.1, we derive

$$\|\mathrm{E}(q,u_1)\|_{L^2_{m+1-\ell}(\Sigma(S')_{\tau'})} \le D_{m,(\mathrm{E}.39)}e^{-\delta_1 S'}. \tag{E.39}$$

Therefore by Lemma B.1

$$\text{(E.40)}\quad \begin{aligned}&\left\|\left(\mathrm{Pal}^{u_1}_{u_1^{\rho_0'}} - \mathrm{Pal}^{u_1}_{q}\circ \mathrm{Pal}^{q}_{u_1^{\rho_0'}}\right)(W)\right\|_{L^2_{m+1-\ell}(\Sigma(S')_{\tau'})}\\ &\le D_{m,(\mathrm{E}.40)}e^{-\delta_1 S'}\|W\|_{L^2_{m+1-\ell}(\Sigma(S')_{\tau'})}\end{aligned}$$

for $W \in \Gamma(\Sigma(S'), (u_1^{\rho_0'})^*TX)$.

Now (E.35), (E.36), (E.38), (E.40) imply

$$\text{(E.41)}\quad \begin{aligned}&\left\|\nabla^n_\rho \frac{\partial^\ell}{\partial T^\ell}\left(\mathfrak{P}_1((\Delta p^\rho_{T,(\kappa)})^{\mathrm{Pal}}) - \mathfrak{P}_1'((\Delta p^\rho_{T,(\kappa)})^{\mathrm{Pal}})\right)\right\|_{L^2_{m+1-\ell}(\Sigma(S')_{\tau'})}\\ &\le (C_{m,(\mathrm{E}.41)}e^{-\delta_1 T} + D_{m,(\mathrm{E}.41)}e^{-\delta_1 S'})\mu^{\kappa-1}e^{-\delta_1 T}.\end{aligned}$$

By taking weighted sum of the square of (E.41), Sublemma E.5 follows easily. □

PROOF OF SUBLEMMA E.6. We take and fix ρ_0' and prove this estimate under the assumption that $d(\rho, \rho_0') \le e^{-\delta_1 T}$.

We put

$$(\Delta q^\rho_{T,(\kappa)})' = (\mathrm{Pal}^q_{u_1^{\rho_0'}} \circ \mathrm{Pal}^{u_1^{\rho_0'}}_{p_0^{\rho_0'}} \circ \mathrm{Pal}^{p_0^{\rho_0'}}_{p_0} \circ \mathrm{Pal}^{p_0}_{p^\rho_{T,(\kappa-1)}})(\Delta p^\rho_{T,(\kappa)}) \in \Gamma(\Sigma(S'), T_qX)$$

and

$$\Delta q^\rho_{T,(\kappa)} = (\mathrm{Pal}^q_{p_0^{\rho_0'}} \circ \mathrm{Pal}^{p_0^{\rho_0'}}_{p_0} \circ \mathrm{Pal}^{p_0}_{p^\rho_{T,(\kappa-1)}})(\Delta p^\rho_{T,(\kappa)}) \in T_qX.$$

(See Figure E.2.)

Note $(\Delta q^\rho_{T,(\kappa)})'$ is $z \in \Sigma(S')$ dependent but $\Delta q^\rho_{T,(\kappa)}$ is a (T,ρ) dependent family of constant maps on $\Sigma(S')$.

By (E.36) we obtain:

$$\text{(E.42)}\quad \begin{aligned}&\left\|\nabla^n_\rho \frac{\partial^\ell}{\partial T^\ell}\left((\mathrm{Pal}^q_{u^\rho_{T,(\kappa-1)}} \circ \mathrm{Pal}^{u^\rho_{T,(\kappa-1)}}_{p^\rho_{T,(\kappa-1)}})(\Delta p^\rho_{T,(\kappa)})) - (\Delta q^\rho_{T,(\kappa)})'\right)\right\|_{L^2_{m+1-\ell}(\Sigma(S'))}\\ &\le C_{m,(\mathrm{E}.42)}\mu^{\kappa-1}e^{-2\delta_1 T}.\end{aligned}$$

We use Lemma B.1 and

$$\text{(E.43)}\quad \left\|\mathrm{E}(p_0^{\rho_0'}, u_1^{\rho_0'})\right\|_{L^2_{m+1-\ell}(\Sigma(S'))} \le D_{m,(\mathrm{E}.43)}e^{-\delta_1 S'}$$

to derive

$$\text{(E.44)}\quad \begin{aligned}&\left\|\nabla^n_\rho \frac{\partial^\ell}{\partial T^\ell}\left(\Delta q^\rho_{T,(\kappa)} - (\Delta q^\rho_{T,(\kappa)})'\right)\right\|_{L^2_{m+1-\ell}(\Sigma(S'))}\\ &\le D_{m,(\mathrm{E}.44)}\mu^{\kappa-1}e^{-\delta_1(T+S')}.\end{aligned}$$

On the other hand, (E.17) and Lemma E.3 imply

(E.45)

$$\begin{aligned}&\Big\|\nabla^n_\rho \frac{\partial^\ell}{\partial T^\ell}\Big(\mathrm{E}(u_1, u^\rho_{T,(\kappa)}) - \mathrm{E}(u_1, u^\rho_{T,(\kappa-1)})\\ &- (\mathrm{Pal}^{u_1}_q \circ \mathrm{Pal}^q_{u^\rho_{T,(\kappa-1)}} \circ \mathrm{Pal}^{u^\rho_{T,(\kappa-1)}}_{p^\rho_{T,(\kappa-1)}})(\Delta p^\rho_{T,(\kappa)})\Big)\Big\|_{L^2_{m+1-\ell,\delta}([0,5T+1]_{\tau'}\times[0,1]\subset\Sigma_T)}\\ &\le D_{m,(\mathrm{E}.45)}\mu^{\kappa-1}e^{-\delta_1 T}.\end{aligned}$$

By (E.42), (E.44) and (E.45) we obtain

$$\begin{aligned}
&\Big\|\nabla_\rho^n \frac{\partial^\ell}{\partial T^\ell}\Big(\mathrm{E}(u_1, u^\rho_{T,(\kappa)}) - \mathrm{E}(u_1, u^\rho_{T,(\kappa-1)}) \\
&\qquad\qquad - \mathrm{Pal}_q^{u_1}(\Delta q^\rho_{T,(\kappa)})\Big)\Big\|_{L^2_{m+1-\ell,\delta}([0,5T+1]_{\tau'}\times[0,1]\subset\Sigma_T)} \\
&\le D_{m,(\mathrm{E}.46)}\mu^{\kappa-1}e^{-\delta_1 T}.
\end{aligned} \tag{E.46}$$

Using Lemma E.3 we can also show

$$\begin{aligned}
&\Big\|\nabla_\rho^n \frac{\partial^\ell}{\partial T^\ell}\Big(\mathrm{E}(u_1, u^\rho_{T,(\kappa)}) - \mathrm{E}(u_1, u^\rho_{T,(\kappa-1)}) \\
&\qquad - (\mathrm{Pal}_q^{u_1}\circ \mathrm{Pal}^q_{u^\rho_{T,(\kappa-1)}} \circ \mathrm{Pal}^{u^\rho_{T,(\kappa-1)}}_{p^\rho_{T,(\kappa-1)}})(\Delta p^\rho_{T,(\kappa)})\Big)\Big\|_{L^2_{m+1-\ell,\delta}(\Sigma(S')_{\tau'}\subset\Sigma_T)} \\
&\le C_{m,(\mathrm{E}.47)}\mu^{\kappa-1}e^{-2\delta_1 T}.
\end{aligned} \tag{E.47}$$

for $S' \ge T$. In fact $u^\rho_{T,(\kappa-1)}$, $u^\rho_{T,(\kappa)}$, $p^\rho_{T,(\kappa-1)}$ are all close to each other by the order of $e^{-\delta_1 T}$ there, including their (T,ρ) derivatives. Also u_1 and q are close by the order of $e^{-\delta_1 T}$, there. Therefore all the error terms appearing while applying Lemma E.3 are of the order $\mu^{\kappa-1}e^{-2\delta_1 T}$.

(E.42), (E.44) and (E.47) imply

$$\begin{aligned}
&\Big\|\nabla_\rho^n \frac{\partial^\ell}{\partial T^\ell}\Big(\mathrm{E}(u_1, u^\rho_{T,(\kappa)})(0,1/2) - \mathrm{E}(u_1, u^\rho_{T,(\kappa-1)})(0,1/2) \\
&\qquad\qquad - \Delta q^\rho_{T,(\kappa)}\Big)\Big\|_{L^2_{m+1-\ell,\delta}(\Sigma(S')_{\tau'}\subset\Sigma_T)} \\
&\le C_{m,(\mathrm{E}.48)}\mu^{\kappa-1}e^{-2\delta_1 T}.
\end{aligned} \tag{E.48}$$

(E.18) follows easily from (E.46) and (E.48). □

APPENDIX F

Estimate of Parallel transport 3

PROOF OF LEMMA 6.21. The proof is similar to the argument of Appendix D. We put

$$W_0(z) = \mathrm{E}(u_1(z), \hat u^{\rho}_{1,T,(\kappa)})$$

then we estimate $\mathbf{e}'_i(z,0)$ in (D.2) and obtain

$$\begin{aligned} &\left\| \nabla^n_\rho \left(\frac{\partial^\ell}{\partial T^\ell} \right) \mathbf{e}'_i(z,0) \right\|_{L^2_{m+1-\ell}} \\ &\le C_{m,(\mathrm{F}.1)} \left\| \nabla^n_\rho \left(\frac{\partial^\ell}{\partial T^\ell} \right) (W_0) \right\|_{L^2_{m+1-\ell}} \le C'_{m,(\mathrm{F}.1)} e^{-\delta_1 T}. \end{aligned} \tag{F.1}$$

Here the second inequality follows from (6.10) and (6.11).

Then (D.4) and (D.5) imply the required inequality (6.59). □

APPENDIX G

Estimate of T derivative of the error term of non-linear Cauchy-Riemann equation

PROOF OF LEMMA 6.14. We discuss estimate on K_1 only. Estimate on $K_1^{4T-1} \setminus K_1$ is similar. We put $K_1^+ = K_1^1 = K_1 \cup [-5T, -5T+1]_\tau \times [0,1]$.

We use the simplified notation:

$$
\begin{aligned}
&u = u^{\rho}_{1,T,(\kappa-1)}, && V = V^{\rho}_{T,1,(\kappa)}, \\
&\mathcal{P} = (\mathrm{Pal}_1^{(0,1)})^{-1}, && \mathfrak{e} = \sum_{a=0}^{\kappa-1} \mathfrak{e}^{\rho}_{1,T,(a)}.
\end{aligned} \tag{G.1}
$$

where $\mathrm{Pal}_1^{(0,1)}$ is $0,1$ part of the parallel transport $r \to \mathrm{Exp}(u(z), rV)$.

We apply Lemma A.1 to

$$
W^0(z) = \mathrm{E}(u_1(z), u(z)), \qquad V^0(z) = \mathrm{Pal}_{u(z)}^{u_1(z)}(V(z)) \tag{G.2}
$$

and use induction hypothesis. We remark that V^0, W^0, $\mathfrak{e}$ are T dependent. We divide $K_1 \subset \bigcup \Omega_a$ such that u_1^*TX etc. are trivial on Ω_a. (See the beginning of Appendix A.)

We put

$$
\mathfrak{P} = (\mathrm{Pal}_{u_1}^{u})^{(0,1)}.
$$

We then can show:

LEMMA G.1.

$$
\begin{aligned}
&\left\| \nabla_\rho^n \frac{d^\ell}{dT^\ell} \mathfrak{P}^{-1} \int_0^1 ds \int_0^s \frac{d^2}{dr^2} \left(\mathcal{P}\overline{\partial}(\mathrm{Exp}(u, rV)\right) dr \right\|_{L^2_{m-\ell}(K_1)} \\
&\le C_{m,(\mathrm{G.3})} \left(\sum_{n'=0}^{n} \sum_{j=0}^{\ell} \left\| \nabla_\rho^{n'} \frac{d^j}{dT^j} V \right\|_{L^2_{m-j+1}(K_1^+)} \right)^2 \\
&\le C'_{m,(\mathrm{G.3})} e^{-2\delta_1 T} \mu^{2(\kappa-1)}.
\end{aligned} \tag{G.3}
$$

PROOF. For simplicity of notation we consider the case $n = 0$. In the case $n \neq 0$ (that is, the case when the ρ derivative is included) the proof is the same.

We take ℓ-th T derivative of (A.7) and find that

$$\begin{aligned}
&\frac{d^\ell}{dT^\ell}(\text{A.7})\\
&= \sum_{\substack{ij,\ell_1,\ell_2,\ell_3\\ \ell_1+\ell_2+\ell_3=\ell}} \frac{\partial^{\ell_1} V_i^0}{\partial T^{\ell_1}} \frac{\partial^{\ell_2} V_j^0}{\partial T^{\ell_2}} \mathfrak{G}_{ij}^{\ell_1,\ell_2,\ell_3}\Big(r, z, W^0, \dots, \frac{\partial^{\ell_3} W^0}{\partial T^{\ell_3}}, V^0, \dots, \frac{\partial^{\ell_3} V^0}{\partial T^{\ell_3}},\\
&\qquad\qquad \frac{\partial W^0}{\partial x}, \dots, \frac{\partial^{\ell_3+1} W^0}{\partial x \partial T^{\ell_3}}, \frac{\partial W^0}{\partial y}, \dots, \frac{\partial^{\ell_3+1} W^0}{\partial y \partial T^{\ell_3}},\\
&\qquad\qquad \frac{\partial V^0}{\partial x}, \dots, \frac{\partial^{\ell_3+1} V^0}{\partial x \partial T^{\ell_3}}, \frac{\partial V^0}{\partial y}, \dots, \frac{\partial^{\ell_3+1} V^0}{\partial y \partial T^{\ell_3}}\Big)\\
&\quad + \sum_{\substack{ij,\ell_1,\ell_2,\ell_3\\ \ell_1+\ell_2+\ell_3=\ell}} \frac{\partial^{\ell_1} V_i^0}{\partial T^{\ell_1}} \frac{\partial^{\ell_2+1} V_j^0}{\partial x \partial T^{\ell_2}} \mathfrak{G}_{ij}^{x,\ell_1,\ell_2,\ell_3}\left(r, z, W^0, \dots, \frac{\partial^{\ell_3} W^0}{\partial T^{\ell_3}}, V^0, \dots, \frac{\partial^{\ell_3} V^0}{\partial T^{\ell_3}}\right)\\
&\quad + \sum_{\substack{ij,\ell_1,\ell_2,\ell_3\\ \ell_1+\ell_2+\ell_3=\ell}} \frac{\partial^{\ell_1} V_i^0}{\partial T^{\ell_1}} \frac{\partial^{\ell_2+1} V_j^0}{\partial y \partial T^{\ell_2}} \mathfrak{G}_{ij}^{y,\ell_1,\ell_2,\ell_3}\left(r, z, W^0, \dots, \frac{\partial^{\ell_3} W^0}{\partial T^{\ell_3}}, V^0, \dots, \frac{\partial^{\ell_3} V^0}{\partial T^{\ell_3}}\right),
\end{aligned}$$

where $\mathfrak{G}_{ij}^{\ell_1,\ell_2,\ell_3}$, $\mathfrak{G}_{ij}^{x,\ell_1,\ell_2,\ell_3}$ and $\mathfrak{G}_{ij}^{y,\ell_1,\ell_2,\ell_3}$ are smooth maps of the variables in the parentheses. Note V^0 there, for example, means $V^0(z) \in \mathbb{R}^n$.

By induction hypothesis (on ℓ), that is an estimate of $\frac{\partial V_i^0}{\partial x}$, $\frac{\partial V_i^0}{\partial y}$, $\frac{\partial W_i^0}{\partial x}$, $\frac{\partial W_i^0}{\partial y}$ and their ℓ'-derivatives (for $\ell' \le \ell$) with respect to T, we can show that $L^2_{m-\ell_3}(\Omega_a)$ norms of the maps

$$\mathfrak{G}_{ij}^{\ell_1,\ell_2,\ell_3}(r, z, \dots, \frac{\partial^{\ell_3+1} V^0}{\partial y \partial T^{\ell_3}}), \mathfrak{G}_{ij}^{x,\ell_1,\ell_2,\ell_3}(r, z, \dots, \frac{\partial^{\ell_3} V^0}{\partial T^{\ell_3}})$$

and

$$\mathfrak{G}_{ij}^{y,\ell_1,\ell_2,\ell_3}(r, z, \dots, \frac{\partial^{\ell_3} V^0}{\partial T^{\ell_3}})$$

are bounded.

Therefore

$$\text{(G.4)} \qquad \left\| \frac{d^\ell}{dT^\ell}(\text{A.7}) \right\|_{L^2_{m-\ell}(\Omega_a)} \le C_{m,(\text{G.4})} \left(\sum_{j=0}^{\ell} \left\| \frac{d^j}{dT^j} V \right\|_{L^2_{m-j+1}(K_1^+)} \right)^2.$$

Taking the sum over a, we obtain the first inequality.

The second inequality is the consequence of induction hypothesis. □

LEMMA G.2. *We have the next two inequalities.*

$$\begin{aligned}
&\left\| \nabla_\rho^n \frac{d^\ell}{dT^\ell} \mathfrak{P}^{-1}(D_u\mathcal{E}_1)(\mathcal{P}Q, V) \right\|_{L^2_{m-\ell}(K_1)}\\
\text{(G.5)} \qquad &\le C_{m,(\text{G.5})} \left(\sum_{n'=0}^{n} \sum_{\ell'=0}^{\ell} \left\| \nabla_\rho^{n'} \frac{d^{\ell'}}{dT^{\ell'}} Q \right\|_{L^2_{m-\ell'}(K_1^+)} \right)\\
&\quad \times \left(\sum_{n'=0}^{n} \sum_{\ell'=0}^{\ell} \left\| \nabla_\rho^{n'} \frac{d^{\ell'}}{dT^{\ell'}} V \right\|_{L^2_{m-\ell'}(K_1^+)} \right).
\end{aligned}$$

$$\begin{aligned}&\left\|\nabla_\rho^n \frac{d^\ell}{dT^\ell}\mathfrak{P}^{-1}\int_0^1 ds\int_0^s \left(\frac{\partial}{\partial r}\right)^2 \left(\mathcal{P}\circ \Pi^\perp_{\mathcal{E}_1(\mathrm{Exp}(u,sV))}\circ \mathcal{P}^{-1}\right) dr\right\|_{L^2_m(K_1)} \\ &\le C_{m,(\mathrm{G.6})}\left(\sum_{n'=0}^{n}\sum_{\ell'=0}^{\ell}\left\|\nabla_\rho^{n'}\frac{d^{\ell'}}{dT^{\ell'}}V\right\|_{L^2_{m-\ell'}(K_1^+)}\right)^2.\end{aligned} \tag{G.6}$$

Here Q, V are T,ρ dependent sections of $u^*TX\times\Lambda^{0,1}$, u^*TX, respectively.

PROOF. We prove this lemma in a similar way to (G.3) and also to Lemma E.3 as follows. For simplicity of formula we consider the case $n=0$.

We divide $K_1\subset\bigcup\Omega_a$ as above. We define W^0, V^0 as in (G.2) and apply the calculation in Appendix D to $\mathbf{e}'_i(z,r)$ and $\mathbf{e}_i(z,r)$.

By the smoothness of $\hat{\mathbf{e}}_i$ and (D.2) the next inequality holds for $\ell>0$.

$$\begin{aligned}&\left\|\left(\frac{d}{dT}\right)^\ell \mathbf{e}'_i\right\|_{L^2_{m-\ell}(\Omega_a)} \\ &\le C_{m,(\mathrm{G.7})}\sum_{k=1}^{\ell}\left(\left\|\frac{d^kV^0}{dT^k}\right\|_{L^2_{m-k}(K_1^+)}+\left\|\frac{d^kW^0}{dT^k}\right\|_{L^2_{m-k}(K_1^!)}\right) \\ &\le C'_{m,(\mathrm{G.7})}e^{-\delta_1T}.\end{aligned} \tag{G.7}$$

We also have the next formula for $\ell\ge 0$.

$$\left\|\left(\frac{d}{dT}\right)^\ell \frac{d}{dr}\mathbf{e}'_i\right\|_{L^2_{m-\ell}(\Omega_a)} \le C_{m,(\mathrm{G.8})}\left(\sum_{k=0}^{\ell}\left\|\left(\frac{d}{dT}\right)^k V^0\right\|_{L^2_{m-k}(K_1^+)}\right). \tag{G.8}$$

Then using (D.4) and (D.5) we can prove

$$\begin{aligned}&\left\|\left(\frac{d}{dT}\right)^\ell \mathbf{e}_i\right\|_{L^2_{m-\ell}(\Omega_a)} \le C_{m,(\mathrm{G.9})}e^{-\delta_1T}, \\ &\left\|\left(\frac{d}{dT}\right)^\ell \frac{d}{dr}\mathbf{e}_i\right\|_{L^2_{m-\ell}(\Omega_a)} \le C'_{m,(\mathrm{G.9})}\left(\sum_{k=0}^{\ell}\left\|\left(\frac{d}{dT}\right)^k V^0\right\|_{L^2_{m-k}(K_1^+)}\right),\end{aligned} \tag{G.9}$$

by induction on i. Using this formula and (D.7), (D.8) it is easy to show (G.5). The proof of (G.6) is similar. □

We use the inequalities (G.3), (G.5) in place of (C.3), (C.7), respectively, for our estimate of T-derivatives. We use (G.6) to estimate the T derivatives of the third term of (C.5). We can then prove

$$\left\|\frac{d^\ell}{dT^\ell}(\mathfrak{P}')^{-1}\left(\overline{\partial}(\mathrm{Exp}(u,V))-\mathcal{P}^{-1}\mathfrak{e}\right)\right\|_{L^2_m(K_1)} \le C_{m,(\mathrm{G.10})}e^{-\delta T}\mu^{\kappa-1} \tag{G.10}$$

in the same way as (C.6). (Here we put $\mathfrak{P}'=(\mathrm{Pal}_{u_1}^{\mathrm{Exp}(u,V)})^{(0,1)}$.)

We then deduce

$$\begin{aligned}\Big\Vert \frac{d^\ell}{dT^\ell}(\mathfrak{P}')^{-1}\Big(\Pi^\perp_{\mathcal{E}_1(\mathrm{Exp}(u,V))}\overline{\partial}(\mathrm{Exp}(u,V)) - \mathcal{P}^{-1}\Pi^\perp_{\mathcal{E}_1(u)}(\mathcal{P}\overline{\partial}(\mathrm{Exp}(u,V)) \\ + \mathcal{P}^{-1}(D_u\mathcal{E}_1)(\mathcal{P}\mathfrak{e},V)\Big)\Big\Vert_{L^2_m(K_1^+)} \\ \le C_{m,(\mathrm{G}.11)}e^{-2\delta T}\mu^{\kappa-1},\end{aligned} \tag{G.11}$$

in a similar way as (5.51).

The rest of the proof of Lemma 6.14 is a straightforward modification of the proof of (C.10) in the same way as above and so is omitted. □

PROOF OF (6.32). We take ρ_0' and prove the inequality for ρ with $d(\rho,\rho_0') \le e^{-\delta_1 T}$. Let $\Omega = [S', S'+1]_{\tau'} \times [0,1] \subset [T, 9T]_{\tau'} \times [0,1]$. We consider $u_1' = p_0^{\rho_0'}$ (constant map) and

$$\begin{aligned} W^0(z) &= \mathrm{E}(p_0^{\rho_0'}, u^\rho_{T,(\kappa-1)}(z)), \\ V^0(z) &= \Big(\mathrm{Pal}^{p_0^{\rho_0'}}_{u^\rho_{T,(\kappa-1)}(z))} \circ \mathrm{Pal}^{u^\rho_{T,(\kappa-1)}(z))}_{p^\rho_{T,(\kappa-1)}}\Big)(\Delta p^\rho_{T,(\kappa)}).\end{aligned} \tag{G.12}$$

We then apply Lemma A.1. Note

$$v_r = \mathrm{Exp}(u^\rho_{T,(\kappa-1)}, r(\Delta p^\rho_{T,(\kappa)})^{\mathrm{Pal}}).$$

We put

$$\mathcal{P}'\mathcal{P} = ((\mathrm{Pal}^{v_0}_{p_0^{\rho_0'}})^{(0,1)})^{-1} \circ ((\mathrm{Pal}^{v_r}_{v_0})^{(0,1)})^{-1}.$$

LEMMA G.3.

$$\Big\Vert \nabla^n_\rho \frac{\partial^\ell}{\partial T^\ell}\frac{\partial}{\partial r}\Big|_{r=0} \mathcal{P}'\mathcal{P}\overline{\partial}v_r\Big\Vert_{L^2_{m-\ell}(\Omega)} \le C_{m,(\mathrm{G}.13)}\mu^{\kappa-1}e^{-2T\delta_1}. \tag{G.13}$$

PROOF. By differentiating (A.3) once by r we obtain

$$\begin{aligned}\frac{\partial}{\partial r}\Big|_{r=0} \mathcal{P}'\mathcal{P}\overline{\partial}v_r &= \sum_i V_i^0\mathfrak{F}_i\left(z, W^0, \frac{\partial W^0}{\partial x}, \frac{\partial W^0}{\partial y}\right) \\ &\quad + \sum_i \frac{\partial V_i^0}{\partial x}\mathfrak{F}_i^x(z,W^0) + \sum_i \frac{\partial V_i^0}{\partial y}\mathfrak{F}_i^y(z,W^0).\end{aligned} \tag{G.14}$$

Note

$$\begin{aligned}&\Big\Vert \nabla^n_\rho \frac{\partial^\ell}{\partial T^\ell}\Big(\mathrm{Pal}^{p_0^{\rho_0'}}_{u^\rho_{T,(\kappa-1)}(z))} \circ \mathrm{Pal}^{u^\rho_{T,(\kappa-1)}(z))}_{p^\rho_{T,(\kappa-1)}} - \mathrm{Pal}^{p_0^{\rho_0'}}_{p^\rho_{T,(\kappa-1)}}\Big)\Big\Vert_{L^2_{m-\ell}(\Omega)} \\ &\le C_{m,(\mathrm{G}.15)}\mu^{\kappa-1}e^{-\delta_1 T}\end{aligned} \tag{G.15}$$

follows from (6.11) and (6.9) in the same way as in Appendix B. Using also the fact that $\mathrm{Pal}^{p_0^{\rho_0'}}_{p^\rho_{T,(\kappa-1)}}(\Delta p^\rho_{T,(\kappa)})$ is constant on Ω, we can use Lemma B.1 to estimate the 2nd and 3rd terms of (G.14) by $C\mu^{\kappa-1}e^{-2\delta_1 T}$.

We also remark that when we substitute $W^0 = 0$ then (G.14) vanishes. In fact if $W^0 = 0$ then $v_r(z) = \mathrm{Exp}(p_0^{\rho_0'}, r\mathrm{Pal}^{p_0^{\rho_0'}}_{p^\rho_{T,(\kappa-1)}}(\Delta p^\rho_{T,(\kappa)}))$, which is a constant map. So $\overline{\partial}v_r = 0$.

Therefore

$$\begin{aligned}
&\left\| \frac{\partial}{\partial r}\bigg|_{r=0} \mathcal{P}'\mathcal{P}\overline{\partial} v_r \right\|_{L^2_m(\Omega)} - C_{m,(\mathrm{G.15})}\mu^{\kappa-1}e^{-2\delta_1 T} \\
&\le C_{m,(\mathrm{G.16})}\|V_0\|_{L^2_m(\Omega)} \left(\|W_0\|_{L^2_m(\Omega)} + \left\|\frac{\partial W^0}{\partial x}\right\|_{L^2_m(\Omega)} + \left\|\frac{\partial W^0}{\partial y}\right\|_{L^2_m(\Omega)} \right) \\
&\le C'_{m,(\mathrm{G.16})}\|V_0\|_{L^2_m(\Omega)}\|W_0\|_{L^2_{m+1}(\Omega)} \\
&\le C''_{m,(\mathrm{G.16})}\mu^{\kappa-1}e^{-2T\delta_1}.
\end{aligned} \tag{G.16}$$

Here the last inequality follows from (6.9) and (6.11). (Note we are working on $[T, 9T]_{\tau'} \times [0,1]$.)

The case when T and ρ derivatives are included is similar. □

We put

$$\mathcal{P}(\mathcal{P}')^{-1} = ((\mathrm{Pal}_{u_1}^{v_0})^{(0,1)})^{-1} \circ (\mathrm{Pal}^{v_0}_{p_0^{\rho'_0}})^{(0,1)}.$$

Here $u_1 : \Sigma_1 \to X$ is the map we start with. (It is different from u'_1.) $v_0 = u^{\rho}_{T,(\kappa-1)}$. This is a (ρ, T) parameterized family of sections of the bundle

$$\mathrm{Hom}(T_{p_0^{\rho'_0}}X, u_1^*TX)\otimes\Lambda^{01}$$

on our domain Ω.

Let $W \in T_{p_0^{\rho'_0}}X\otimes\Lambda^{01}$. Then, by (6.11) we can show the inequality

$$\begin{aligned}
&\|\nabla^n_{\rho}\frac{\partial^{\ell}}{\partial T^{\ell}}(\mathcal{P}(\mathcal{P}')^{-1} - ((\mathrm{Pal}_{u_1}^{p_0^{\rho'_0}})^{(0,1)})^{-1})(W)\|_{L^2_{m-\ell}(\Omega)} \\
&\le C_{m,(\mathrm{G.17})}e^{-\delta_1 T}|W|
\end{aligned} \tag{G.17}$$

by using Lemma B.1.

Note $((\mathrm{Pal}_{u_1}^{p_0^{\rho'_0}})^{(0,1)})^{-1}$ is (T, ρ) independent and is bounded.

Therefore by putting $W = \mathcal{P}'\mathcal{P}\overline{\partial} v_r$, the formulae $\mathcal{P}(\mathcal{P}')^{-1} \circ \mathcal{P}'\mathcal{P} = \mathcal{P}\mathcal{P}$, (G.13) and (G.17) imply

$$\left\| \nabla^n_{\rho}\frac{\partial^{\ell}}{\partial T^{\ell}}\frac{\partial}{\partial r}\bigg|_{r=0} \mathcal{P}\mathcal{P}\overline{\partial} v_r \right\|_{L^2_m(\Omega)} \le C_{m,(\mathrm{G.18})}\mu^{\kappa-1}e^{-2T\delta_1}. \tag{G.18}$$

where

$$\mathcal{P}\mathcal{P} = ((\mathrm{Pal}_{u_1}^{v_0})^{(0,1)})^{-1} \circ ((\mathrm{Pal}_{v_0}^{v_r})^{(0,1)})^{-1}.$$

(6.32) follows immediately from this formula. In fact

$$\frac{\partial}{\partial r}\bigg|_{r=0} \mathcal{P}\mathcal{P}\overline{\partial} v_r = ((\mathrm{Pal}_{u_1}^{u^{\rho}_{T,(\kappa-1)}})^{(0,1)})^{-1}(D_{u^{\rho}_{T,(\kappa-1)}}\overline{\partial})((\Delta^{\rho}_{T,(\kappa)})^{\mathrm{Pal}}).$$

□

APPENDIX H

Proof of Lemma 8.4

Let Γ_+ be a finite group and Γ be its index 2 subgroup. We assume that there exists $\tau \in \Gamma_+$ of order 2 such that $\Gamma_+ = \Gamma \cup \tau\Gamma$.

A smooth action of Γ_+ to a complex manifold M is said to be *holomorphic* if all the action of elements of Γ is holomorphic and τ's action is by an anti-holomorphic involution.

LEMMA H.1. *Let $p \in M$ be a fixed point of a holomorphic action of Γ_+ then there exists a complex coordinate of M at p such that the Γ_+ action is linear in this coordinate.*

PROOF. We fix a coordinate of M at p. Then Γ_+ induces a holomorphic action ρ on $U \subset \mathbb{C}^n$ such that 0 is a fixed point. We consider the Γ_+ action on the tangent space at 0 which we denote by ρ_0. For $t \in (0,1]$ we define the action ρ_t by

$$\rho_t(g,x) = \rho(g,tx)/t.$$

Together with ρ_0 it defines a smooth one parameter family of holomorphic actions. Consider the map

$$g \mapsto \left.\frac{d}{dt}\right|_{t=t_0} \rho_t(g, \rho_{t_0}(g^{-1},x)) \in \mathrm{Hol}(U, T\mathbb{C}^n).$$

Here $\mathrm{Hol}(U,TX)$ is the vector space of holomorphic vector fields on U. For each t_0 the map defines a 1 cycle of the complex $C^*(\Gamma_+, \mathrm{Hol}(U, T\mathbb{C}^n))$ which calculates the group cohomology with local coefficients induced by the conjugate action associated to ρ_{t_0}. (Note the conjugate action by the anti-holomorphic involution τ preserves the set of holomorphic vector fields.)

Using the fact that Γ_+ is a finite group we can show that the first cohomology group vanishes. In fact the Eilenberg-MacLane space $K(\Gamma_+,1)$ has a finite cover $\widetilde{K(\Gamma_+,1)}$ which is contractible. All the local systems on $K(\Gamma_+,1)$ pulled back to $\widetilde{K(\Gamma_+,1)}$ are trivial and the first cohomology groups of the pulled back are trivial. Since $\mathbb{Q}$ is contained in the coefficient ring of our local system, the Gysin map induces an isomorphism between cohomologies of $K(\Gamma_+,1)$ and of $\widetilde{K(\Gamma_+,1)}$.

So there exists a one parameter family of holomorphic vector fields V_t such that

$$\left.\frac{d}{dt}\right|_{t=t_0} \rho_t(g, \rho_{t_0}(g^{-1},x)) = \rho_{t_0}(g,\cdot)_* V_t - V_t.$$

By integrating V_t we obtain a biholomorphic map which interpolates ρ_0 and ρ. The lemma follows. □

LEMMA H.2. *Let M, N be complex manifolds on which Γ_+ has holomorphic actions. Let $p \in M$ and $q \in N$ be fixed points of Γ_+. Assume $F : M \to N$ is a*

Γ_+ equivariant holomorphic map such that $F(p) = q$ and $d_pF : T_pM \to T_qN$ is surjective.

We decompose $T_pM = T_qN \oplus V$ as a Γ_+ complex vector space such that d_pF is identity on T_qN and is 0 on V.

Then there exists a Γ_+ equivariant biholomorphic map

$$\varphi : B_\epsilon(T_qN) \times B_\epsilon(V) \to M$$

sending $(0,0)$ to p such that $F \circ \varphi$ is constant in the V direction.

PROOF. Using Lemma H.1 we may assume that M, N are open subsets W_1, W_2 of $\mathbb{C}^M$, $\mathbb{C}^N$ and the Γ_+ action is linear. We may assume $p = 0$, $q = 0$. We take two Γ_+ invariant complex linear subspaces U_1, U_2 of $\mathbb{C}^M$ such that $U_1 \oplus U_2 = \mathbb{C}^M$, $(d_0F)(U_1) = 0$ and d_0F induces an isomorphism between U_2 and $\mathbb{C}^N$. We define $G : W_1 \to U_1 \times W_2$ by $G(x + y) = (x, F(x + y))$, where $x \in U_1, y \in U_2$. By the inverse mapping theorem G has a local inverse φ. Then $F \circ \varphi$ is the projection to the W_2 factor. Therefore φ gives the required coordinate. (The Γ_+ equivariance and holomorphicity of φ are obvious from construction.) $\square$

PROOF OF LEMMA 8.4. Applying Lemma H.2 to $\mathcal{C}^{\rm cl}(\mathcal{V}) \to \mathcal{V}$ we can prove Lemma 8.4 easily. $\square$

Bibliography

[Ab] Mohammed Abouzaid, *Framed bordism and Lagrangian embeddings of exotic spheres*, Ann. of Math. (2) **175** (2012), no. 1, 71–185, DOI 10.4007/annals.2012.175.1.4. MR2874640

[BF] Jean-Michel Bismut and Daniel S. Freed, *The analysis of elliptic families. I. Metrics and connections on determinant bundles*, Comm. Math. Phys. **106** (1986), no. 1, 159–176. MR853982

[DM] P. Deligne and D. Mumford, *The irreducibility of the space of curves of given genus*, Inst. Hautes Études Sci. Publ. Math. **36** (1969), 75–109. MR262240

[D1] S. K. Donaldson, *An application of gauge theory to four-dimensional topology*, J. Differential Geom. **18** (1983), no. 2, 279–315. MR710056

[D2] S. K. Donaldson, *Connections, cohomology and the intersection forms of 4-manifolds*, J. Differential Geom. **24** (1986), no. 3, 275–341. MR868974

[DK] S. K. Donaldson and P. B. Kronheimer, *The geometry of four-manifolds*, Oxford Mathematical Monographs, The Clarendon Press, Oxford University Press, New York, 1990. Oxford Science Publications. MR1079726

[FU] Daniel S. Freed and Karen K. Uhlenbeck, *Instantons and four-manifolds*, Mathematical Sciences Research Institute Publications, vol. 1, Springer-Verlag, New York, 1984, DOI 10.1007/978-1-4684-0258-2. MR757358

[Fu] Kenji Fukaya, *Floer homology of connected sum of homology 3-spheres*, Topology **35** (1996), no. 1, 89–136, DOI 10.1016/0040-9383(95)00009-7. MR1367277

[FOOO] Kenji Fukaya, Yong-Geun Oh, Hiroshi Ohta, and Kaoru Ono, *Lagrangian intersection Floer theory: anomaly and obstruction. Part I*, AMS/IP Studies in Advanced Mathematics, vol. 46, American Mathematical Society, Providence, RI; International Press, Somerville, MA, 2009, DOI 10.1090/amsip/046.1. MR2553465

[FOOO1] Kenji Fukaya, Yong-Geun Oh, Hiroshi Ohta, and Kaoru Ono, *Lagrangian intersection Floer theory: anomaly and obstruction. Part II*, AMS/IP Studies in Advanced Mathematics, vol. 46.2, American Mathematical Society, Providence, RI; International Press, Somerville, MA, 2009, DOI 10.1090/amsip/046.2. MR2548482

[FOOO2] K. Fukaya, Y.-G. Oh, H. Ohta, and K. Ono, *Technical details on Kuranishi structure and virtual fundamental chain*, Preprint, `arXiv:1209.4410`, 2013.

[FOOO3] Kenji Fukaya, Yong-Geun Oh, Hiroshi Ohta, and Kaoru Ono, *Construction of Kuranishi structures on the moduli spaces of pseudo holomorphic disks: I*, Surveys in differential geometry 2017. Celebrating the 50th anniversary of the Journal of Differential Geometry, Surv. Differ. Geom., vol. 22, Int. Press, Somerville, MA, 2018, pp. 133–190. MR3838117

[FOOO4] Kenji Fukaya, Yong-Geun Oh, Hiroshi Ohta, and Kaoru Ono, *Kuranishi structures and virtual fundamental chains*, Springer Monographs in Mathematics, Springer, Singapore, [2020] ©2020, DOI 10.1007/978-981-15-5562-6. MR4179586

[FOn] Kenji Fukaya and Kaoru Ono, *Arnold conjecture and Gromov-Witten invariant*, Topology **38** (1999), no. 5, 933–1048, DOI 10.1016/S0040-9383(98)00042-1. MR1688434

[HWZ] H. Hofer, K. Wysocki, and E. Zehnder, *A general Fredholm theory. I. A splicing-based differential geometry*, J. Eur. Math. Soc. (JEMS) **9** (2007), no. 4, 841–876, DOI 10.4171/JEMS/99. MR2341834

[Kl] Wilhelm Klingenberg, *Lectures on closed geodesics*, Grundlehren der Mathematischen Wissenschaften, Vol. 230, Springer-Verlag, Berlin-New York, 1978. MR0478069

[McSa] Dusa McDuff and Dietmar Salamon, *J-holomorphic curves and quantum cohomology*, University Lecture Series, vol. 6, American Mathematical Society, Providence, RI, 1994, DOI 10.1090/ulect/006. MR1286255

[Mr] Tomasz Stanislaw Mrowka, *A local Mayer-Vietoris principle for Yang-Mills moduli spaces*, ProQuest LLC, Ann Arbor, MI, 1988. Thesis (Ph.D.)–University of California, Berkeley. MR2637291

[Oh1] Yong-Geun Oh, *Removal of boundary singularities of pseudo-holomorphic curves with Lagrangian boundary conditions*, Comm. Pure Appl. Math. **45** (1992), no. 1, 121–139, DOI 10.1002/cpa.3160450106. MR1135926

[Oh2] Yong-Geun Oh, *Symplectic topology and Floer homology. Vol. 2: Floer homology and its applications*, New Mathematical Monographs, vol. 29, Cambridge University Press, Cambridge, 2015. MR3524783

[Q] D. Quillen, *Determinants of Cauchy-Riemann operators on Riemann surfaces* (Russian), Funktsional. Anal. i Prilozhen. **19** (1985), no. 1, 37–41, 96. MR783704

[LM] Robert B. Lockhart and Robert C. McOwen, *Elliptic differential operators on noncompact manifolds*, Ann. Scuola Norm. Sup. Pisa Cl. Sci. (4) **12** (1985), no. 3, 409–447. MR837256

[Ye] Rugang Ye, *Gromov's compactness theorem for pseudo holomorphic curves*, Trans. Amer. Math. Soc. **342** (1994), no. 2, 671–694, DOI 10.2307/2154647. MR1176088

List of notations

Index

Editorial Information

To be published in the *Memoirs*, a paper must be correct, new, nontrivial, and significant. Further, it must be well written and of interest to a substantial number of mathematicians. Piecemeal results, such as an inconclusive step toward an unproved major theorem or a minor variation on a known result, are in general not acceptable for publication.

Papers appearing in *Memoirs* are generally at least 80 and not more than 200 published pages in length. Papers less than 80 or more than 200 published pages require the approval of the Managing Editor of the Transactions/Memoirs Editorial Board. Published pages are the same size as those generated in the style files provided for $\mathcal{A}\mathcal{M}\mathcal{S}$-LaTeX.

Information on the backlog for this journal can be found on the AMS website starting from `http://www.ams.org/memo`.

A Consent to Publish is required before we can begin processing your paper. After a paper is accepted for publication, the Providence office will send a Consent to Publish and Copyright Agreement to all authors of the paper. By submitting a paper to the *Memoirs*, authors certify that the results have not been submitted to nor are they under consideration for publication by another journal, conference proceedings, or similar publication.

Information for Authors

Memoirs is an author-prepared publication. Once formatted for print and on-line publication, articles will be published as is with the addition of AMS-prepared frontmatter and backmatter. Articles are not copyedited; however, confirmation copy will be sent to the authors.

Initial submission. The AMS uses Centralized Manuscript Processing for initial submissions. Authors should submit a PDF file using the Initial Manuscript Submission form found at `www.ams.org/submission/memo`, or send one copy of the manuscript to the following address: Centralized Manuscript Processing, MEMOIRS OF THE AMS, 201 Charles Street, Providence, RI 02904-2294 USA. If a paper copy is being forwarded to the AMS, indicate that it is for *Memoirs* and include the name of the corresponding author, contact information such as email address or mailing address, and the name of an appropriate Editor to review the paper (see the list of Editors below).

The paper must contain a *descriptive title* and an *abstract* that summarizes the article in language suitable for workers in the general field (algebra, analysis, etc.). The *descriptive title* should be short, but informative; useless or vague phrases such as "some remarks about" or "concerning" should be avoided. The *abstract* should be at least one complete sentence, and at most 300 words. Included with the footnotes to the paper should be the 2020 *Mathematics Subject Classification* representing the primary and secondary subjects of the article. The classifications are accessible from `www.ams.org/msc/`. The Mathematics Subject Classification footnote may be followed by a list of *key words and phrases* describing the subject matter of the article and taken from it. Journal abbreviations used in bibliographies are listed in the latest *Mathematical Reviews* annual index. The series abbreviations are also accessible from `www.ams.org/msnhtml/serials.pdf`. To help in preparing and verifying references, the AMS offers MR Lookup, a Reference Tool for Linking, at `www.ams.org/mrlookup/`.

Electronically prepared manuscripts. The AMS encourages electronically prepared manuscripts, with a strong preference for $\mathcal{A}\mathcal{M}\mathcal{S}$-LaTeX. To this end, the Society has prepared $\mathcal{A}\mathcal{M}\mathcal{S}$-LaTeX author packages for each AMS publication. Author packages include instructions for preparing electronic manuscripts, samples, and a style file that generates the particular design specifications of that publication series.

Authors may retrieve an author package for *Memoirs of the AMS* from `www.ams.org/journals/memo/memoauthorpac.html`. The *AMS Author Handbook* is available in PDF format from the author package link. The author package can also be obtained free of charge by sending email to `tech-support@ams.org` or from the Publication Division,

American Mathematical Society, 201 Charles St., Providence, RI 02904-2294, USA. When requesting an author package, please specify the publication in which your paper will appear. Please be sure to include your complete mailing address.

After acceptance. The source files for the final version of the electronic manuscript should be sent to the Providence office immediately after the paper has been accepted for publication. The author should also submit a PDF of the final version of the paper to the editor, who will forward a copy to the Providence office.

Accepted electronically prepared files can be submitted via the web at `www.ams.org/submit-book-journal/`, sent via FTP, or sent on CD to the Electronic Prepress Department, American Mathematical Society, 201 Charles Street, Providence, RI 02904-2294 USA. TEX source files and graphic files can be transferred over the Internet by FTP to the Internet node `ftp.ams.org` (130.44.1.100). When sending a manuscript electronically via CD, please be sure to include a message indicating that the paper is for the *Memoirs*.

Electronic graphics. Comprehensive instructions on preparing graphics are available at `www.ams.org/authors/journals.html`. A few of the major requirements are given here.

Submit files for graphics as EPS (Encapsulated PostScript) files. This includes graphics originated via a graphics application as well as scanned photographs or other computer-generated images. If this is not possible, TIFF files are acceptable as long as they can be opened in Adobe Photoshop or Illustrator.

Authors using graphics packages for the creation of electronic art should also avoid the use of any lines thinner than 0.5 points in width. Many graphics packages allow the user to specify a "hairline" for a very thin line. Hairlines often look acceptable when proofed on a typical laser printer. However, when produced on a high-resolution laser imagesetter, hairlines become nearly invisible and will be lost entirely in the final printing process.

Screens should be set to values between 15% and 85%. Screens which fall outside of this range are too light or too dark to print correctly. Variations of screens within a graphic should be no less than 10%.

Any graphics created in color will be rendered in grayscale for the printed version unless color printing is authorized by the Managing Editor and the Publisher. In general, color graphics will appear in color in the online version.

Inquiries. Any inquiries concerning a paper that has been accepted for publication should be sent to `memo-query@ams.org` or directly to the Electronic Prepress Department, American Mathematical Society, 201 Charles St., Providence, RI 02904-2294 USA.

Editors

SELECTED PUBLISHED TITLES IN THIS SERIES

1486 **M. Gekhtman, M. Shapiro, and A. Vainshtein,** A Plethora of Cluster Structures on GL_n, 2024

1485 **David Ayala, Aaron Mazel-Gee, and Nick Rozenblyum,** Stratified Noncommutative Geometry, 2024

1484 **Brandon Doherty, Krzysztof Kapulkin, Zachery Lindsey, and Christian Sattler,** Cubical Models of $(\infty, 1)$-Categories, 2024

1483 **Masao Oi,** Simple Supercuspidal L-Packets of Quasi-Split Classical Groups, 2024

1482 **David Burns and Daniel Macias Castillo,** On Refined Conjectures of Birch and Swinnerton-Dyer Type for Hasse–Weil–Artin L-Series, 2024

1481 **Sonja Cox, Martin Hutzenthaler, and Arnulf Jentzen,** Local Lipschitz Continuity in the Initial Value and Strong Completeness for Nonlinear Stochastic Differential Equations, 2024

1480 **Nicolas Burq and Laurent Thomann,** Almost Sure Scattering for the One Dimensional Nonlinear Schrödinger Equation, 2024

1479 **Eva Elduque, Christian Geske, Moisés Herradón Cueto, Laurenţiu G. Maxim, and Botong Wang,** Mixed Hodge Structures on Alexander Modules, 2024

1478 **Qi Chen, Dongyi Wei, and Zhifei Zhang,** Transition Threshold for the 3D Couette Flow in a Finite Channel, 2024

1477 **Vincent Guirardel, Gilbert Levitt, and Rizos Sklinos,** Towers and the First-Order Theories of Hyperbolic Groups, 2024

1476 **Gaëtan Borot, Vincent Bouchard, Nitin K. Chidambaram, Thomas Creutzig, and Dmitry Noshchenko,** Higher Airy Structures, $\mathcal{W}$ Algebras and Topological Recursion, 2024

1475 **Sam Chow and Niclas Technau,** Littlewood and Duffin–Schaeffer-Type Problems in Diophantine Approximation, 2024

1474 **Christian Blohmann and Alan Weinstein,** Hamiltonian Lie Algebroids, 2024

1473 **Dihua Jiang, Zhilin Luo, and Lei Zhang,** Harmonic Analysis and Gamma Functions on Symplectic Groups, 2024

1472 **Thomas Creutzig, Shashank Kanade, and Robert McRae,** Tensor Categories for Vertex Operator Superalgebra Extensions, 2024

1471 **Roberto Feola and Filippo Giuliani,** Quasi-Periodic Traveling Waves on an Infinitely Deep Perfect Fluid Under Gravity, 2024

1470 **Maxwell Stolarski,** Curvature Blow-up in Doubly-warped Product Metrics Evolving by Ricci Flow, 2024

1469 **Steve Boyer, Cameron McA. Gordon, and Xingru Zhang,** Dehn Fillings of Knot Manifolds Containing Essential Twice-Punctured Tori, 2024

1468 **Alice Hedenlund and John Rognes,** A Multiplicative Tate Spectral Sequence for Compact Lie Group Actions, 2024

1467 **Qi Lü and Xu Zhang,** Optimal Feedback for Stochastic Linear Quadratic Control and Backward Stochastic Riccati Equations in Infinite Dimensions, 2024

1466 **Martin W. Liebeck, Gary M. Seitz, and Donna M. Testerman,** Multiplicity-free Representations of Algebraic Groups, 2024

1465 **Tobias Rossmann and Christopher Voll,** Groups, Graphs, and Hypergraphs: Average Sizes of Kernels of Generic Matrices with Support Constraints, 2024

1464 **Jens Marklof and Andreas Strömbergsson,** Kinetic Theory for the Low-Density Lorentz Gas, 2024

1463 **You Qi and Joshua Sussan,** p-DG Cyclotomic nilHecke Algebras II, 2024

For a complete list of titles in this series, visit the
AMS Bookstore at **www.ams.org/bookstore/memoseries/**.